List of titles

Already published

Cell Differentiation	J.M. Ashworth
Biochemical Genetics	R.A. Woods
Functions of Biological Membranes	M. Davies
Cellular Development	D. Garrod
Brain Biochemistry	H.S. Bachelard
Immunochemistry	M.W. Steward
The Selectivity of Drugs	A. Albert
Biomechanics	R. McN. Alexander
Molecular Virology	T.H. Pennington, D.A. Ritchie
Hormone Action	A. Malkinson
Cellular Recognition	M.F. Greaves
Cytogenetics of Man and other Animals	A. McDermott
RNA Biosynthesis	R.H. Burdon
Protein Biosynthesis	A.E. Smith
Biological Energy Conservation	C. Jones
Control of Enzyme Activity	P. Cohen
Metabolic Regulation	R. Denton, C.I. Pogson
Plant Cytogenetics	D.M. Moore
Population Genetics	L.M. Cook
Insect Biochemistry	H.H. Rees
A Biochemical Approach to Nutrition	R.A. Freedland, S. Briggs
Enzyme Kinetics	P.C. Engel
Polysaccharide Shapes	D.A. Rees
Transport Phenomena in Plants	D.A. Baker
Cellular Degradative Processes	R.J. Dean
Human Genetics	J.H. Edwards

In preparation

The Cell Cycle	S. Shall
Microbial Metabolism	H. Dalton, R.R. Eady
Bacterial Taxonomy	D. Jones, M. Goodfellow
Molecular Evolution	W. Fitch
Metal Ions in Biology	P.M. Harrison, R. Hoare
Muscle	R.M. Simmons
Xenobiotics	D.V. Parke
Biochemical Systematics	J.B. Harbourne
Biochemical Pharmacology	B.A. Callingham
Biological Oscillations	A. Robertson
Photobiology	K. Poff
Membrane Assembly	J. Haslam

OUTLINE STUDIES IN BIOLOGY

Editor's Foreword

The student of biological science in his final years as an undergraduate and his first years as a graduate is expected to gain some familiarity with current research at the frontiers of his discipline. New research work is published in a perplexing diversity of publications and is inevitably concerned with the minutiae of the subject. The sheer number of research journals and papers also causes confusion and difficulties of assimilation. Review articles usually presuppose a background knowledge of the field and are inevitably rather restricted in scope. There is thus a need for short but authoritative introductions to those areas of modern biological research which are either not dealt with in standard introductory textbooks or are not dealt with in sufficient detail to enable the student to go on from them to read scholarly reviews with profit. This series of books is designed to satisfy this need. The authors have been asked to produce a brief outline of their subject assuming that their readers will have read and remembered much of a standard introductory textbook of biology. This outline then sets out to provide by building on this basis, the conceptual framework within which modern research work is progressing and aims to give the reader an indication of the problems, both conceptual and practical, which must be overcome if progress is to be maintained. We hope that students will go on to read the more detailed reviews and articles to which reference is made with a greater insight and understanding of how they fit into the overall scheme of modern research effort and may thus be helped to choose where to make their own contribution to this effort. These books are guidebooks, not textbooks. Modern research pays scant regard for the academic divisions into which biological teaching and introductory textbooks must, to a certain extent, be divided. We have thus concentrated in this series on providing guides to those areas which fall between, or which involve, several different academic disciplines. It is here that the gap between the textbook and the research paper is widest and where the need for guidance is greatest. In so doing we hope to have extended or supplemented but not supplanted main texts, and to have given students assistance in seeing how modern biological research is progressing, while at the same time providing a foundation for self help in the achievement of successful examination results.

J.M. Ashworth, Professor of Biology, University of Essex.

Transport Phenomena in Plants

D. A. Baker

Reader in Plant Physiology,
University of Sussex

LONDON
CHAPMAN AND HALL

A Halsted Press Book
John Wiley & Sons, New York

First published in 1978
by Chapman and Hall Ltd
11 New Fetter Lane, London EC4P 4EE
© *1978 D.A. Baker*
Typeset by C. Josée Utteridge-Faivre
Printed in Great Britain at the
University Printing House, Cambridge

ISBN 0 412 15360 2

Distributed in the U.S.A.
by Halsted Press, a Division of
John Wiley & Sons, Inc., New York

Library of Congress Cataloging in Publication Data

Baker, Denis A.
    Transport phenomena in plants.

    (Outlines studies in biology)
    Bibliography: p.
    Includes index.
    1. Plant translocation. I. Title.
QK871.B28   581.1'1   77-27084
ISBN 0-470-26288-5

Contents

1 Introduction

Plants, in addition to their role as primary synthesizers of organic compounds, have evolved as selective accumulators of inorganic nutrients from the earth's crust. This ability to mine the physical environment is restricted to green plants and some microorganisms, other life forms being directly or indirectly dependent on this process for their supply of mineral nutrients. The initial accumulation of ions by plants is often spatially separated from the photosynthetic parts, necessitating the transport to these parts of the inorganic solutes thus acquired. The requirement for energy-rich materials by the accumulation process is provided by a transport in the opposite direction of organic solutes from the photosynthetic areas.

These transport phenomena in plants have been studied at the cellular level, the tissue level, and the whole plant level. The basic problems of analysing the driving forces and the supply of energy for solute transport remain the same for all systems, but the method of approach and the type of results obtained vary widely with the experimental material employed, reflecting the variation of the solute transporting properties which have selectively evolved in response to both internal and external environmental pressures.

It is assumed in this study guide that the reader is sufficiently familiar with plant structure and function to know the location and the role of the major transport processes, and now wishes to discover more about the mechanisms involved. In many instances a satisfactory answer cannot be given at present, the transport of solutes in plants being an area in which our understanding is limited, due to the complexity of the systems studied. In certain areas, such as ion transport in giant algal cells, considerable progress has been made enabling the process to be partially characterized, whereas in other areas, such as the translocation of solutes in the phloem, the basic problem of mechanism remains unresolved and is the subject of considerable controversy.

The unresolved problem of phloem transport has been given more space than some other important topics here, reflecting both its major role within the plant and the renewed emphasis on phloem functioning in recent research studies. Where structure and function have a pronounced interdependence some structural information is provided, although in the case of phloem transport the detailed structural information required to resolve the mechanism is unfortunately not available.

Where it has been possible to analyse the result of these transport processes, in terms of the internal ionic composition, some intriguing features have resulted. In the data presented in Table 1.1 it can be seen that the cytoplasm and phloem sap from diverse plant materials have a similar

Table 1.1 Ionic composition of sea water, animal, and plant materials (mol m^{-3})

Material	K$^+$	Na$^+$	Ca^{2+}	Mg^{2+}	Cl$^-$
Sea water	10	470	11	53	540
Mammalian plasma	4	140	5	2	100
Mammalian erythrocytes	150	6	3	5	80
Nitella cytoplasm	125	5	–	–	36
Nitella vacuole	80	28	16	17	136
Onion cytoplasm (root)	100	9	10	–	30
Onion vacuole (root)	83	44	1	–	24
Ricinus phloem sap	90	7	3	10	15
Ricinus xylem sap	6	1	10	4	1

ionic composition to that of mammalian erythrocytes, particularly the levels of K$^+$ and Na$^+$. The mammalian cell composition is maintained in the relatively constant internal environment of the blood plasma, which has a similar ionic composition to sea water. Most higher plants live and grow in environments in which the ionic composition of the soil solution is very different from that of sea water and can therefore achieve their ionic balance only following complex selective mechanisms. In these terms plant cells have become truly terrestrial. In the following chapters some of the transport mechanisms which have evolved as a result of this terrestrial environment are discussed.

In addition to the specific books listed at the end of each chapter and the individual papers cited in the text, readers may find the following introductory and general books useful.

Bibliography

Gauch, H.G. (1972), *Inorganic Plant Nutrition*, Dowden, Hutchinson and Ross Inc., Stroudsburg, Pa.
A sourcebook on plant nutrition with over 2600 references.
Kramer, P.J. (1969), *Plant and Soil Water Relationships*, McGraw-Hill, New York.
Useful integration of ion uptake studies with those on water absorption.
Meidner, H. and Sheriff, D.W. (1976), *Water and Plants*, Blackie; Glasgow.
A useful introductory text on water movement in the plant incorporating the biophysical aspects of the process.
Nobel, P.S. (1974), *Introduction to Biophysical Plant Physiology*, Freeman, San Francisco.
Biophysical and physicochemical approach to ion uptake and translocation. Suitable for advanced students.
Richardson, M. (1975), *Translocation in Plants*, 2nd edition, Edward Arnold, London.
A short, very simplified account of translocation processes.
Sutcliffe, J.F. and Baker, D.A. (1974), *Plants and Mineral Salts*, Edward Arnold, London.
A short introductory text on mineral nutrition, ion uptake and distribution.
Wardlaw, I.F. and Passioura, J.B. (eds) (1976), *Transport and Transfer Processes in Plants*, Academic Press, New York; San Francisco, London
Symposium proceedings which integrates and reviews a broad spectrum of research on the movement of substances in plants.

2 Solute transport at the cellular level

2.1 Driving forces

The driving forces for solute transport at the membrane level in plants are the same as those in other biological membrane systems. However, the presence of a cell wall and of a vacuole in plant cells are features which add considerably to the technical problems of measuring the various parameters. The basic transport equations (see below) may be applied to transport across plant cell membranes, but the application of such an approach to the cells and tissues of higher plants is severely limited. Often the cells are too small to allow any accurate electrophysiological measurements to be made and much of the basic investigation has so far been made on the transport properties of large algal coenocytes (see Section 2.2).

For uncharged solutes the driving force is the gradient of chemical potential only, while for ions the driving force has an additional component arising from the electrical potential differences (p.d.). If the concentration, or more correctly activity, of an ion is known on either side of a membrane it is possible to predict the magnitude of the p.d. which will maintain an equilibrium state across that membrane. This is given by the Nernst equation:

$$E_{N_j} = \frac{RT}{z_j F} \ln \frac{c_j^0}{c_j^i} \tag{2.1}$$

where E_{N_j} is the Nernst potential in mV for ion species j, R the gas constant, T the absolute temperature, z_j the valency (with sign), F the Faraday constant, and c_j^0/c_j^i the ratio of the concentrations.

The Nernst equation describes the equilibrium condition when the tendency for an ion to move down its chemical potential gradient in one direction is balanced by the tendency to move down its electrical potential gradient in the opposite direction.

The calculated Nernst potential E_{N_j} may be compared with the measured potential across the membrane E_M, indicating whether or not a particular ion species is in passive equilibrium across that membrane. If $E_M = E_{N_j}$, a passive equilibrium situation exists, while if $E_M \neq E_{N_j}$ then energy must be expended to maintain an equilibrium. The difference between E_M and E_{N_j}, ΔE_j is the minimum energy required, the sign of ΔE_j indicating the direction of the net passive driving force which must be opposed by a metabolically-driven active transport if equilibrium is to be maintained:

	ΔE_j value	Direction of passive driving force
Cations	+	i $\longrightarrow$ o
	−	o $\longrightarrow$ i
Anions	+	o $\longrightarrow$ i
	−	i $\longrightarrow$ o

The Nernst equation is applicable only when the ions are in flux equilibrium conditions across the membrane. For non-equilibrium conditions active transport may be identified by the flux ratio for an ionic species across a membrane, measured using suitable radioisotopes and following the initial influx and efflux to give the resultant net flux:

$$\text{net flux}\,(\phi_j) = \text{influx}\,(\phi_j^{oi}) - \text{efflux}\,(\phi_j^{io})$$

The relationship between the influx and efflux ratio is given by the Ussing-Teorell equation:

$$\frac{\phi_j^{oi}}{\phi_j^{io}} = \frac{c_j^0}{c_j^i \exp\,(z_j\,FE_M/RT)}\,. \qquad (2.2)$$

When $\phi_j^{oi} = \phi_j^{io}$, the above equation reduces to the Nernst equation (2.1). Thus when $E_M = E_{N_j}$, no net passive flux of ion j will occur and no energy need be expended in moving the ion from one side of the membrane to the other. When Equation (2.2) is not satisfied, that is when the measured flux ratio differs from the predicted value, ion species j is not moving passively and active transport is indicated, or in some cases the ion is not moving independently of other fluxes. Such interdependent fluxes are describable by irreversible thermodynamics [see 11, 50, 76].

The above criteria for active transport tell us nothing directly of the rate of transport of ions across a membrane, a process which is dependent on a number of factors, in particular the permeability of a membrane to a specific ion. This value, P_j, is given by:

$$P_j = \frac{u_j\,K_j\,RT}{\delta} \qquad (2.3)$$

where u_j is the mobility, K_j the water: membrane partition coefficient, R and T as before and δ the membrane thickness. Mobilities, solubilities and membrane thickness cannot be measured with any great accuracy in biological systems but P_j is relatively easy to estimate from tracer experiments. Expressed in units of m s^{-1}, P_j is the sum of the permeabilities of a number of pathways which may be followed by an ion crossing a membrane, and is therefore subject to some environmental modification.

In some plant cells, particularly algae, the total ionic flux is mainly due to movements of K^+, Na^+ and Cl^-, although H^+ and OH^- fluxes may also be considerable. The resultant diffusion potential across the membrane arises as a result of the different mobilities of these ions under conditions of electrical neutrality which is maintained across the membrane. Considering he major physiologically important ions, K^+, Na^+, and Cl^-,

$$E_M = \frac{RT}{F} \ln \left(\frac{P_K \, c_K^0 + P_{Na} \, c_{Na}^0 + P_{Cl} \, c_{Cl}^i}{P_K \, c_K^i + P_{Na} \, c_{Na}^i + P_{Cl} \, c_{Cl}^0} \right). \tag{2.4}$$

This is the Goldman or Hodgkin-Katz equation which is widely used in interpreting membrane electropotential differences. It predicts that E_M is determined by the passively moving ions. However, in virtually all biological membranes the active transport of some of the ion species involved may also contribute to E_M, if the active transport involves the movement of a charged ion-carrier complex on one or both of its journeys across the membrane. Such a process is termed electrogenic and requires an additional term in the above equation such that

$$E_M = E_{eq} + E_X \tag{2.5}$$

where E_{eq} is a diffusion potential and E_X an additive electrogenic mechanism. In most biological membranes $P_K \approx P_{Cl} > P_{Na}$, and the Goldman equation often gives a similar result to the Nernst equation (2.1) for K^+ and Cl^-. A difference between calculated and predicted membrane potentials does not always indicate an electrogenic pump and additional evidence is required before electrogenicity can be unequivocally claimed. Evidence of electrogenicity is provided by the very rapid depolarization of a membrane with metabolic inhibitors, which often reduce the value of E_M close to that of the predicted Goldman value. When the inhibitor is removed the electrogenic pump commences once again and the cell returns to its resting potential (Fig. 2.1).

Measurements of the electrical conductance of plant cell membranes can provide information on their structure and organization and also indicate the manner in which ions cross them. When an ion species is at electrochemical potential equilibrium a flux, ϕ_j, will contribute a partial ion conductance, g_j, such that

$$g_j = \frac{z_j^2 F^2}{RT} \phi_j. \tag{2.6}$$

For example a ϕ_K of 10 nmol $m^{-2} s^{-1}$ will contribute 38 mmho m^{-2} partial conductance at 20°C, and its reciprocal, the resistance, R_K, would be 260×10^4 kΩm^2. Estimates of ϕ_j or R_j obtained from Eqn. (2.6) may be compared with direct electrical measurements obtained by passing a known current across the membrane. In giant algal cells g_K values of 40 mmho m^{-2} have been calculated for the plasma membrane, whereas measured values give 500–1000 m mho m^{-2} [62]. This observation has two possible explanations, either the K^+ does not carry all the current, or the ion movement is not independent of the movement of other ions. Considering the first possibility, ϕ_{Na}, ϕ_{Cl} and ϕ_{Ca} would not contribute more than a few additional mmho m^{-2}. However, it is possible that ϕ_H may contribute a large part of the conductance in algal cells and probable that ϕ_K is not totally independent of other fluxes.

Current passing through a membrane will, according to its direction of

11

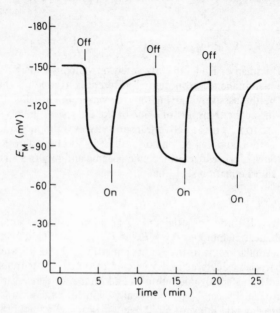

Fig. 2.1 Depolarization-repolarization cycles in the cell potential of a pea root epidermal cell. Respiratory inhibition is by CO, which attaches to haem iron in the dark, while in the light normal O_2 attachment proceeds [6].

flow, either depolarize or hyperpolarize the resting potential of the membrane. Hyperpolarization cannot proceed beyond E_M values of about 300 mV due to the occurrence of the 'punch-through' phenomenon (Fig. 2.2). The membrane behaves in a manner which suggests a semi-conductor composed of alternating layers of opposite fixed change between which there is a depletion layer. As E_M increases, the depletion layer increases in width until it extends to the outer limits of the membrane at which point 'punch-through' occurs and a sudden increase in current is observed.

2.2 Carriers and pumps

The permeability of biological membranes to solute molecules P_j, is known to decrease with increasing polarity of the permeant. This is due primarily to the low partition coefficient K_j of polar solutes, which reflects the high lipid content of cell membranes. Many solutes, particularly ions, must therefore have their movement across the plasma membrane facilitated in some way if they are to accumulate within the cell at observed concentra-tions (see p. 8). Some form of carrier molecule is envisaged, which is soluble in the membrane, combines with the permeant and transports it across the membrane. When such movement is down the gradient of chemical or electrochemical potential the carrier process may be regarded as a facilitated diffusion and the carrier may be 'passive'. However, when solute movement is against the potential gradient an 'active' process is

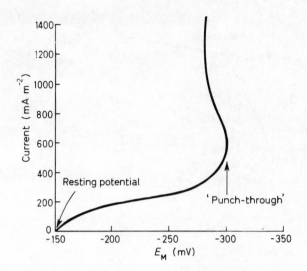

Fig. 2.2 The effect of an increased current on the membrane potential difference, E_M. The resting potential was -150 mV. 'Punch-through' takes place when $E_M = -300$ mV.

invoked, work being done by the carrier to move a solute in the thermodynamically uphill direction.

It has been observed by a large number of investigators [77, 132] that the uptake of ions by plant tissues appears to obey a relationship analogous to the Michaelis-Menten equation used in analysing the kinetics of enzyme reactions. Assuming that ions S, bind reversibly to specific carriers C, in the membrane, the proposed model is

$$S + C \underset{k_{-1}}{\overset{k_1}{\rightleftharpoons}} SC \underset{k_{-2}}{\overset{k_2}{\rightleftharpoons}} C + S \tag{2.7}$$

where k_1, k_{-1}, k_2, k_{-2} are the rate constants of the reaction. Assuming there is little or no counterflow from inside of the membrane to outside $(k_{-2} \cong 0)$ then the velocity of ion uptake v at ion concentration $[S]$ in the external medium is given by

$$v = \frac{[S] \times V_{max}}{K_s + [S]} \tag{2.8}$$

where V_{max} is the maximum rate of uptake when all the carrier is saturated, while K_s is a constant, characteristic of a particular ion crossing a specific membrane, expressed in units of concentration (mol m^{-3}).

The reciprocal of Eqn. (1.8) gives a linear relationship

$$\frac{1}{v} = \frac{K_s}{V_{max}[S]} + \frac{1}{V_{max}} \tag{2.9}$$

13

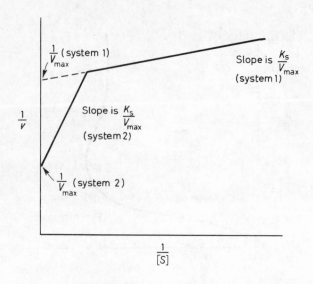

Fig. 2.3 Double reciprocal plot of ion uptake by plant tissue which often yields two lines from which different K_s and V_{max} values may be obtained.

and when $1/v$ is plotted against $1/[S]$ a straight line with an intercept of $1/V_{max}$ on the ordinate and a slope of K_s/V_{max} is obtained. This enables values for V_{max} and K_s to be calculated and the affinity of a carrier for an ion may thus be characterized. Uptake of an ion by plant tissues often yields two or more lines when the above relationship is plotted, from which different K_s and V_{max} values may be obtained (Fig. 2.3). This has been interpreted as evidence for separate carriers within the membrane, each with a different affinity for a particular ion [35]. Often two such carriers are postulated for a particular ionic species; one functional at low concentrations (< 0.5 mol m^{-3}) with a low K_s and thus a high affinity, termed system 1, the other functional at higher concentrations (> 0.5 mol m^{-3}) with a higher K_s and lower affinity, termed system II. The system II uptake isotherm is frequently irregular and may be further subdivided [89].

The location of these two uptake systems has been hotly debated, some investigators suggesting that system I is on the plasma membrane, system II on the tonoplast, and they are thus arranged in series [73]. Others subscribe to the view that both mechanisms are located on the plasma membrane, arranged in parallel [35] (Fig. 2.4).

Alternative models have been proposed to explain the dual isotherms without invoking the concept of two or more carrier sites. Only one uptake may be involved, the dual isotherms reflecting either allosteric structural changes in the membrane [89], the membrane behaving as a semi-conductor [117] or a direct contribution of E_M [45]. Invagination of the plasma membrane on which ions are selectively bound has been considered and the dual isotherms interpreted in terms of the delivery of ions to different cytoplasmic compartments [10].

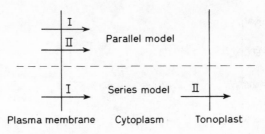

Fig. 2.4

Some investigators have cast doubt on the usefulness of the application of Michaelis-Menten kinetics to ion uptake by plant tissues (see [132]). Ion absorption isotherms which are the result of active processes do not always follow the Michaelis-Menten formalism and passive ion movement, particularly if carrier-mediated, may sometimes manifest saturation kinetics. It is therefore necessary to approach results obtained from this 'carrier' type of analysis with a degree of caution.

Conceptually the above carrier model differs markedly from the 'pumps' which have been postulated to operate in maintaining the ionic composition within giant algal and other plant cells. The major difference is that in the carrier model all movement into and out of the cell is considered carrier-mediated, whereas 'pumps' are proposed to counteract a passive ion movement or leakage in the opposite direction. Simply, the carrier model does not allow for the ready exchange of ions across cell membranes as envisaged by the pump model.

Measurements of the internal and external ion concentration and of the membrane potential have enabled the identification and location of pumps on the plasma membrane and tonoplast of some giant algal coenocytes. In fresh water algae, such as *Nitella* and *Hydrodictyon* active Na^+ efflux (ϕ_{Na}^{io}), K^+ influx (ϕ_{K}^{oi}) and Cl^- influx (ϕ_{Cl}^{oi}) have been identified using the Nernst equation (2.1) and Ussing-Teorell equation (2.2) criteria for active transport [80, 104]. These pumps, located at the plasma membrane, are responsible for the K^+/Na^+ ratio being considerably higher in the cytoplasm than in the external medium of these plant cells.

2.3 Energy sources for active transport

The energy for active transport of solutes may be provided by either ATP hydrolysis or by the electron flow associated with certain redox reactions. Thus in green plants the energy can be provided by respiration or photosynthesis or both. In these processes electrons are fed down an electron transport system of decreasing redox potential resulting in a lowering of the free energy of the electron. At points along this transport system energy is conserved by the phosphorylation of ADP to ATP, the process of photophosphorylation in the chloroplasts and oxidative phosphorylation in mitochondria.

The form of the coupling between electron transport and phosphorylation has been the subject of considerable debate. Originally it was envisaged that the energy transfer occurred through a series of high energy

15

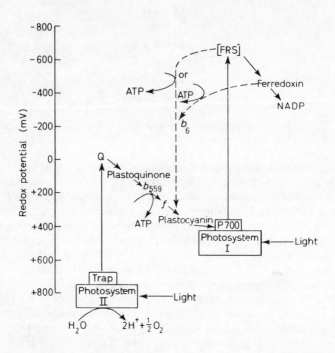

Fig. 2.5 Electron transfer between photosystems I and II. FRS (ferredoxin-reducing substance) may serve as a reductant to ferredoxin [50].

chemical intermediates, the so-called chemical coupling hypothesis. This concept has been challenged and superceded by the chemiosmotic hypothesis which proposes that the electron carriers in the inner mitochondrial and chloroplast membranes have a vectorial arrangement enabling them to generate a gradient of protons in response to the electron flow. The energy derived from this proton gradient, termed the proton motive force, may be coupled to phosphorylation or to provide the potential gradient for the accumulation of ions.

The ion pumps identified in algal coenocytes (see p. 15) require a considerable expenditure of energy. The actual magnitude may be calculated from the value of ΔE_j, 1 mV being equal to 96500 J mol^{-1}. For the Cl$^-$ influx (ϕ_{Cl}^{oi}) in *Nitella* the value of ΔE_j is a staggering 22.9×10^6 J mol^{-1}. As outlined above, this energy may be obtained from either respiration or photosynthesis. In algae such as *Nitella* it has been clearly demonstrated that the major ion pumps, ϕ_{Cl}^{oi}, ϕ_{K}^{oi} and ϕ_{Na}^{io} are light dependent [80, 104] and furthermore they continue to function in the light in the absence of CO_2 [104]. This indicates that it is the 'light' reactions of photosynthesis, in which light energy is converted into chemical energy, which drive these ion pumps.

Two photochemical reactions are normally involved in the light reaction termed photosystems I and II (PS I and II), the proposed arrangement of these photosystems and associated electron transport systems being presented in Fig. 2.5 PS I absorbs light up to 730 nm wavelength while PS II, which requires higher energy quanta (shorter wavelength) absorbs up to 705 nm. Thus red light between 705 and 730 nm will only excite PS I and there will be no reduction of NADP and no oxygen production, but ATP will still be generated by the cyclic photophosphorylation pathway. In *Nitella* illuminated with red light only, ϕ_{Cl}^{oi} is reduced to the dark level while ϕ_K^{oi} and ϕ_{Na}^{io} are unaffected. This result suggests that the cation pumps may be energized by ATP hydrolysis and that the anion pump is independent of this process. In white light, when both PS I and II are operating, uncouplers such as carbonylcyanide m–chlorophenylhydrazone (CCCP) will inhibit the active cation fluxes without affecting ϕ_{Cl}^{oi}, which may even be partially stimulated. These observations indicate a dependence of the cation pumps on ATP formation and a possible link between the anion pump and the electron transport process, since the latter is also stimulated by uncouplers. Dichlorophenyldimentyl urea (DCMU) inhibits PS II by blocking the non-cyclic electron flow between PS II and PS I but does not appear to inhibit cyclic electron flow from PS I alone. This inhibitor reduces the active ϕ_{Cl}^{oi} to the dark level but does not markedly reduce the active cation fluxes.

The above observations with inhibitors indicate that light dependent active cation transport in algal cells may be energized by ATP hydrolysis, whereas the anion transport may be more directly dependent on electron flow. However, such an hypothesis requires non-cyclic electron flow to energize and cyclic electron not to energize the chloride pump in this system, suggesting a possible spatial separation of the transport chains within the chloroplast. The dependence of ϕ_{Cl} on the electron flow does not exist in all organisms studied, in a number of cases, ϕ_{Cl}^{oi} has been found to be sensitive to uncouplers [80, 104], and thus apparently energized by ATP hydrolysis.

The extensive studies of the coupled active $\phi_{Na}^{io} - \phi_K^{oi}$ in mammalian cells, which is driven by a membrane-bound ATPase, has stimulated the search for a similar system in plant cells. The mammalian ATPase is orientated within the membrane and is directly related to the Na^+-K^+ pump. Both enzyme and pump require Na^+ and K^+ together for maximal activity and are specifically inhibited by the cardiac glycoside ouabain. In the alga *Hydrodictyon* both ϕ_K^{oi} and ϕ_{Na}^{io} are similarly inhibited by ouabain and ϕ_{Na}^{io} is dependent on $[K^+]^0$ suggesting that a coupled Na^+-K^+ pump may be present [103]. Cation-stimulated ATPases have been characterized from a variety of higher plant sources, but at present the clear correlations reported in animal cells have not been observed [49]. It may be that with improved techniques for isolating plant cell membranes some specific transport ATPases may be demonstrated, and those pumps energized by ATP hydrolysis understood in greater detail.

A problem arises when considering how a membrane-located pump may be energized by the electron transport system. How can events restricted to the mitochondria and chloroplasts be transmitted to the plasma membrane through an appreciable thickness of highly buffered cytoplasm? One possibility is that proton-containing microvesicles are shunted across the

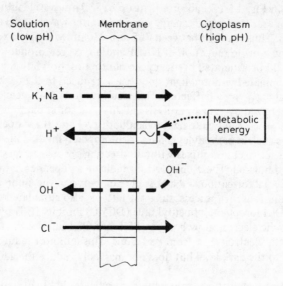

Fig. 2.6 A scheme for net salt transport at the plasma membrane of algal cells. The primary energy-requiring process is ϕ_H^{io}, which is partly balanced by ϕ_K^{oi} and ϕ_{Na}^{oi}. The remainder of the ϕ_H^{io} is balanced by ϕ_{OH}^{io} coupled to an active ϕ_{Cl}^{oi} [111].

cytoplasm to the plasma membrane and that a proton extrusion pump drives the chloride uptake. A model for such a proton-driven ion transport across the plasma membrane has been proposed (Fig. 2.6) and some evidence provided by an increased ϕ_{Cl}^{oi} when the ΔH^+ is increased by lowering the pH of the external solution.

2.4 Sensitive cells

Plants exhibit a number of characteristic movements which are nastic responses, that is they are made in response to stimuli, but are not directionally orientated by the stimulus. Some of these movements are rapid and striking, as in the response of the sensitive plant *Mimosa*, or are slower and rhythmic, as in the nyctinastic or 'sleep' movements observed in the leaves of a number of plants. These variation movements are usually turgor responses involving the rapid movement of water into or out of certain sensitive cells.

The regulation of stomatal aperture is a variation movement of great physiological significance, since both the uptake of CO_2 for photosynthesis and the loss of transpirational water from the plant are controlled by the turgor responses of the two guard cells which form the stomatal pore. Simply, when the guard cells are turgid the pore is open and when they are flaccid the pore is closed. The response of these sensitive guard cells to a variety of external stimuli is relatively rapid, opening or closing of stomata taking a matter of minutes; thus considerable water exchange must take place between the guard and surrounding cells. This water movement is

18

Table 2.1 The amounts of various elements, measured with the electron probe microanalyser, in open and closed guard cells, the guard cell volumes, and the changes in stomatal apertures and osmotic pressure [66]

	Amounts per stoma (10^{-14} mol)			Stomatal aperture (μm)	Guard cell volume per stoma (10^{-15} m^{-3})	Guard cell osmotic pressure from incipient plasmolysis (bars)
	K$^+$	Na$^+$	Cl$^-$			
Open stomata	424	0	22	12	4.8	35
Closed stomata	20	0	0	2	2.6	19
Difference between open and closed stomata	404	0	22	10	2.2	16

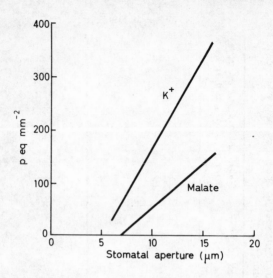

Fig. 2.7 K^+ and malate levels in epidermal strips of *Vicia faba* at various stomatal apertures [2].

considered to be due to solute accumulation in the guard cells. Early worker considered that starch hydrolysis within the guard cells would produce osmotically-active solutes which in turn provide a gradient for water movement. Such a scheme does not require that any solute movement be involved and early evidence that K^+ accumulated in open guard cells [78] was largely ignored until relatively recently when it was postulated that a light-activated K^+ accumulation in the guard cells could play a major role in stomatal opening [43]. Subsequent measurement with the electron probe microanalyser (EPM) [66] and with K^+-sensitive microelectrodes [94] have clearly implicated K^+ as a major solute which is transported during stomatal movements (Table 2.1). Specificity for K^+ is not found in plants with a crassulacean acid metabolism (CAM), where the normal light response is reversed and the stomata open in the dark; in this case Na^+ is reported to be the cation involved [118]. These CAM plants may be induced to fix CO_2 in the dark by the carboxylation of phosphoenol pyruvate (PEP) to form malate under condition of salinity and water stress [130] and the Na^+-dependent stomatal movements are presumably related to this adaptive mechanism.

Although a high specificity has been observed for the cation involved [65] little specificity has been found for the accompanying anion, the opening of stomata on epidermal strips being equally supported by equivalent concentrations of K^+ in solutions of KCl, $K_2 SO_4$, KBr and KNO_3. Even when K^+ was supplied accompanied by anions considered to be non-absorbed (e.g. iminodiacetate or benzenesulphonate) the opening response was equal to that of KCl [102]. In EPM studies of *Vicia faba* guard cells, it was found that only 5% of the K^+ was balanced by inorganic anions [66], indicating that organic anions must be present in large amounts to

Table 2.2 pH and concentrations of di- and monovalent malate (mol m^{-3}) in the guard and epidermal cells of *Commelina communis* [23]

Stomata open	Guard cell	Epidermal cell
pH	5.8	5.1
Malate$^-$	60	32
Malate^{2-}	120	9
Stomata closed		
pH	5.1	5.8
Malate$^-$	32	60
Malate^{2-}	9	120

preserve electroneutrality. In *Zea mays,* approximately 40% of the charge balance is provided by Cl^- [101] but again the remainder must be organic anions. Malate could be the balancing organic anion as it has been found that guard cell malate content increases with stomatal opening (Fig. 2.7) although it is not clear whether the malate is synthesized in the guard cells or transported there from surrounding cells. If the malate is synthesized in the guard cells, CO_2 fixation may not be rapid enough and breakdown of starch and other polysaccharide reserves is the most likely source. However, if the guard cells operated a C_4 pathway of photosynthesis the synthesis of malate via PEP carboxylase could provide the anion balance.

The emphasis has been on K^+ movement with an accompanying anion but it has been pointed out that the anion may have the major role. The concentration of malate in the guard and epidermal cells of *Commelina communis* has been calculated and an interesting switch of valence takes place as the pH moves through the pK value of 5.2 for malate (Table 2.2). At pH $<$ 5.2 malate is monovalent and at pH $>$ 5.2 malate is divalent. Divalent malate is believed to be relatively immobile and a movement of monovalent malate into or out of the guard cells could cause stomata to open and close. It is a well-established fact that pH has an effect on stomatal aperture, high pH values induce opening and low pH values induce closure. At pH 5.1 the malate in the guard cell will switch to the monovalent form and move down a gradient into the surrounding epidermal cells resulting in stomatal closure. At pH $>$ 5.2 in the guard cell the switch to divalent malate will favour the movement of monovalent into the guard cells resulting in stomatal opening. This hypothesis requires that a pH change in the guard cells must always precede stomatal movements.

The diurnal 'sleep' movements of the leaves of many plants also appears to be related to solute movements. Often a specialized group of cells, the *pulvinus,* located at the base of the leaf, is responsible for these movements. It appears that water can move rapidly into or out of motor cells, located on opposite sides of the pulvinus, and the resulting turgor changes cause the leaf to move up or down. As a result, in the light the leaves move down (open out) while in the dark they close up. EPM studies have shown that K^+ accumulates in the lower pulvinal cells, with a corresponding loss from the upper pulvinal cells, and induces turgor changes resulting in upward leaf movement. The K^+ gradient is reversed when downward leaf movement takes place [106]. This solute movement appears to be specific for K^+ and is dependent upon metabolism. Isolated pulvini show a similar response and undergo K^+ redistribution in response to a light stimulus indicating that they are photoreceptive. Experiments indicate that red light (< 730 nm) induces closure movements which can be reversed by far red light (> 730 nm), implicating phytochrome as the photoreceptor [71]. Phytochrome action in this and in many other systems may be simply represented by the following

$$\text{red light} \quad \downarrow \quad \begin{array}{c} \text{Phytochrome (red-light absorbing)} \\[4pt] \text{Phytochrome (far-red light absorbing)} \end{array} \quad \uparrow \quad \text{far-red light}$$

$$\downarrow$$

$$\text{Biological action}$$

It is possible that phytochrome, a membrane-bound pigment, may act through changes in membrane permeability, red light inducing a decreased permeability with greater selective ion transport and far-red light inducing an increased permeability [69].

In many cases the movements induced by excitable cells within plants may be in response to a transmitted stimulus. In the sensitive plant *Mimosa pudica* the plants respond to touch or mechanical shock by folding their leaflets and lowering their leaves. This response may start after 0.1 seconds and be completed within a few seconds, the reaction spreading up and down the plant at rates of 400 to 500 mm s^{-1}. This stimulus may be transmitted by the propagation of a neuroid action potential. In the carnivorous plant *Drosera* the propagation of electrically induced action potentials has been reported at rates of 4 to 10 mm s^{-1} probably by an electrotonic mechanism through plasmodesmata [128]. The response of *Mimosa* is faster and over a greater distance, but may also be transmitted electrotonically by an action potential. However, attempted analogies with the nervous system of animals should be tempered by recorded propagation rates of 4–160 m s^{-1} in myelinated animal nerves, although rates of 1–2 m s^{-1} are normal through the non-myelinated nerves of some invertebrates.

Bibliography

Clarkson, D.T. (1974), *Ion Transport and Cell Structure in Plants*, McGraw-Hill, Maidenhead.
A clear account of ion transport across cell membranes with particular emphasis on Hydrodictyon as a case history. The second part of this book is useful for Chapters 3 and 4 here.

Hall, J.L. and Baker, D.A. (1977), *Membranes and Ion Transport*, Longman, London.
An introductory text on membranes and ion transport in both plant and animal systems.

Hope, A.B. (1971), *Ion Transport and Membranes – a Biological Outline*, Butterworths, London.
An advanced biophysical approach requiring fairly sophisticated mathematical knowledge.

Hope, A.B. and Walker, N.A. (1975), *The Physiology of Giant Algal Cells*, Cambridge University Press, London.
A scholarly account of transport processes in giant algae.

Lüttge, U. and Pitman, M.G. (eds.) (1976), *Encyclopedia of Plant Physiology*
   N.S. Volume 2A, Springer-Verlag, Berlin, Heidelberg, New York.
An excellent multiauthor treatise on ion transport at the cellular level.
Stein, W.D. (1967), *The Movement of Molecules Across Cell Membranes*,
   Academic Press, New York.
*A fine monograph on biological transport. Chapter 3 is particularly useful
but unfortunately most of the examples are of transport in animal
systems.*

3 Symplast and apoplast

3.1 The parallel pathways

There are two possible pathways for solute movement through plant tissues,
the cell wall pathway and the cytoplasmic pathway. Movement may take
place extracellularly through channels in the cell walls, thus bypassing the
protoplasts and obviating the need to cross membrane barriers. Solutes in
these cell-wall channels can be freely exchanged with an external solution
and therefore this pathway is often referred to as the free space, the whole
cell-wall continuity being termed the apoplast or apoplasm. In parallel with
the free-space pathway is a route through the cytoplasm of the protoplasts,
which are linked through the plasmodesmata to form a three-dimensional
cytoplasmic continuity known as the symplast or symplasm. Solutes entering
the symplast must first cross the plasma membrane, imparting a high degree
of selectivity upon this pathway.

The relative importance of these two pathways differs within the various
organs of the plant. In the case of the uptake of ions by the young root the
free space provides a continuum between the external solution and the cells
of the root cortex. Uptake into this cell wall phase is reversible (Fig. 3.1)
and thus its volume may be estimated by washing out the solutes following a
period of uptake, assuming that the concentration of solutes in the free
space is the same as that in the external solution. However, this volume
differs for various solutes and when ions are used the contribution from the
exchange capacity of the cell walls compounds the measurement. The
concept of apparent free space (AFS) has been introduced to allow for this
exchange contribution, but in general it is preferable to express the free
space as a quantity rather than as a volume. The water-extractable fraction
is therefore those solutes which are free in the aqueous phase of the cell
wall, sometimes referred to as the water free space (WFS), and the ion-
exchangeable fraction those ions which are reversibly bound in the electrical
double layer or Donnan phase of the cell wall, the so-called Donnan free
space (DFS).

The external solution can move across the cortex through the free space

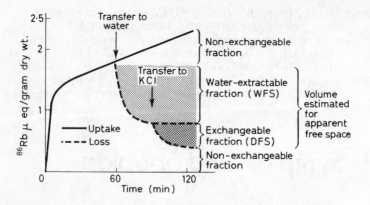

Fig. 3.1 The uptake and loss of ^{86}Rb-labelled K$^+$ by maize roots. The initial rapid uptake may be reversed by washing out the water-extractable and the exchangeable fractions with water and unlabelled KCl respectively. Th show, steady uptake process evident after 15 minutes exchanges less readily.

as far as the endodermis, which provides a barrier to any further radial movement through the apoplast pathway (see Section 3.2). Thus the plasma membrane of the cortical cells is bathed in the external solution and a vast surface area is available for ion uptake into the cortical cytoplasm. The root cortex thereby selectively accumulates ions from the external medium into the cytoplasmic continuity, the symplast.

The ionic content within the symplast is regulated by the selective properties of the uptake mechanism and by selective accumulation of ions across the tonoplast into the vacuole. Whereas high levels of K$^+$ are a feature of the symplast, Na$^+$ is usually preferentially accumulated in the cortical vacuoles and is thus partially removed from the symplast. Some ions within the symplast are utilised in a metabolic capacity, while the majority are moved radially across the root and into the stele. It has been proposed that this radial movement is diffusional, the driving force being the declining gradient of concentration across the root [122] with cytoplasmic streaming aiding this ion movement. Inhibition of cytoplasmic streaming within the root cortical cells inhibits radial ion transport [8]. There is only a very limited longitudinal movement in the cortical symplast, the direction of the driving force being centripetal. Some evidence of a gradient of K$^+$ across the root has been provided using the electron probe microanalyser [74], although work with K$^+$ selective microelectrodes reveals no significant trend [21], as shown in Fig. 3.2. This difference may reflect a high [K$^+$] in the cytoplasm, which is included with the electron probe analyser measurement but not with the K$^+$ microelectrode, which measures only vacuolar activity. A gradient of pH across the root has been reported [20], rising from less than 6.0 at the epidermis to nearly 7.0 at the protoxylem, although this observation is difficult to interpret, as once again it is a vacuolar measurement and it is the cytoplasmic value which is critical here.

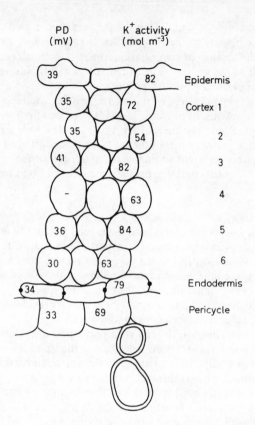

Fig. 3.2 Part of a transverse section through a *Helianthus annuus* root, detailing the cell electrical potentials and the vacuolar K^+ activities [21].

The ions which are transported radially across the root cross the endodermis through the symplast and are ultimately released within the stele for long-distance transport through the xylem. There has been some debate as to the mechanism of ion release within the stele, some investigators advocating a passive, thermodynamically downhill, process [29] while others envisage an active efflux across the plasma membrane of the stelar parenchyma [134]. It appears that stelar cells are capable of accumulating ions when applied through the symplast but are incapable of accumulation when isolated from the cortical tissue [9]. There are data which show that both anions and cations move into the xylem down the gradient of electro-chemical potential [34], consistent with the passive leakage hypothesis. However, some investigators still favour an active efflux mechanism within the stele, using the high accumulative capacity of xylem parenchyma as evidence of a secretory role [74] and inhibition of root pressure exudation by metabolic poisons as support for this hypothesis [96]. As with some other problems in the transport physiology of plants, we await definative experiments to resolve this conflict of interpretation.

25

Ions released into the apoplast of the stele are then moved to the aerial portions of the plant with the transpiration stream through the xylem conduits. In the absence of transpiration, the ions accumulate within the stelar apoplast and lower the water potential, the resultant water movement across the root creating a positive pressure which forces a solution of ions up the xylem. This phenomenon, root pressure, may be readily observed from detopped root systems, or in whole plants as guttation fluid which extrudes from the leaves. Root pressure is often cited as evidence for a radial barrier to the apoplastic pathway within the root, the composition of root pressure exudate being very different to that of the external solution, while a constant electrical p.d. exists between exudate and external solution across the root.

3.2 Radial barriers – the endodermis

The importance of an apoplastic barrier has been referred to in the preceding section when considering the radial transport of ions across the root through the symplast. Most investigators support the view that this barrier is the endodermis, where the lignin and suberin impregnated Casparian band is believed to form the inner limit of the cortical free space (Fig. 3.3). Evidence that the endodermis is a limiting structure has been provided by a number of autoradiographic, EM localization and ion exchange studies, all of which indicate movement through the cell walls only up to the Casparian band, but not beyond. However, there is evidence that at the tip of the root where the endodermis is not fully differentiated and the Casparian band is not developed free space continuity exists through into the xylem [105].

When the Casparian band is developed there is a strong attachment at that point between the plasma membrane of the endodermal cell and the impregnated cell wall, which is not broken even when subjected to plasmolysing conditions. Water and solutes which have crossed the low resistance pathway of the cortical apoplast are obliged to cross the plasma membrane at this point, a feature which results in the root having a similar hydraulic conductivity (L_p) value to that of a single plasma membrane of a cell such as *Nitella*, approximately 0.5×10^{-8} m s^{-1} bar^{-1}. Once across the plasma membrane of the endodermis solutes can continue their radial movement in the symplast through plasmodesmata into the stele, or may be transferred into the stelar apoplast on the inner side of the Casparian band.

In many roots, particularly those of monocots, there is thickening of the inner tangential and radial walls of the endodermal cells as a result of lignin and suberin deposition. Not all cells of the endodermis show this thickening and unthickened cells termed 'passage' cells are often found opposite the xylem archs of the stele. These passage cells do possess a Casparian band however and do eventually become thickened. It is this thickening of endodermal cells which gives the characteristic toothed appearance to the endodermis in transverse sections of the root of these plants (Fig. 3.4). However, the presence of large pits in the inner tangential wall ensures symplastic continuity across the thickened endodermis. Plasmodesmata have been reported at frequencies of 670 000 mm^{-2} in these inner tangential walls even at a distance of 300–400 mm from the root tip [27].

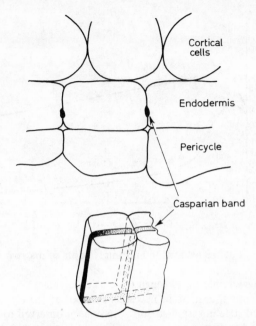

Fig. 3.3 Diagram to show the Casparian band in the endodermis, located in the transverse and radial walls, but not in the tangential walls.

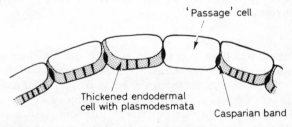

Fig. 3.4 Thickened endodermis with 'passage' cell.

An interesting correlation between the development of endodermal structure and radial ion transport is the observation that K^+ is translocated readily into the stele all along the root whereas Ca^{2+} is discriminated against in the more mature regions of the root (Fig. 3.5). In the development of the endodermis, suberin is deposited on the secondary cell walls and when this is complete the endodermis is only accessible to water and solutes through the plasmodesmata. This development of the endodermis coincides with the reduced ability to translocate Ca^{2+}. Accessibility of the endodermal plasma membrane to the cortical free space thus appears to be a requisite for Ca^{2+} transfer to the stele and it would appear that Ca^{2+} cannot move readily in the symplasm. The corollary of this argument is that K^+, which does not have its radial movement impaired by these developmental changes, is

27

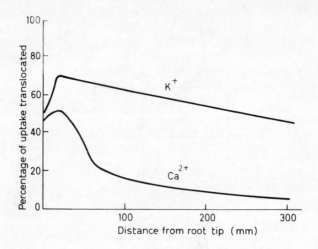

Fig. 3.5 K^+ and Ca^{2+} translocation in primary roots of marrow [26].

delivered to the endodermis through the cortical symplasm [26].

An endodermis appears to be necessary for root pressure exudation, as neither isolated stele nor isolated cortex have been observed to exude [9]. When the stele and cortex are separated the break occurs at the Casparian band depriving both components of an apoplasmic barrier. However, active movement of Cl^- across the symplasm of isolated cortical sleeves has been observed although K^+ and Na^+ show no such active transport moving passively across the cortex [47].

3.2.1 *Water flux across the root*
It would appear that the endodermis of an intact root functions as a radial barrier for movement through the apoplast, all movement into the stele being directed across a membrane at this point. By preventing any back diffusion of ions through the cell walls the endodermis also provides an osmotic barrier across which water moves to drive root pressure exudation, although some investigations have indicated that this osmotic barrier may be located elsewhere within the stele [63].

There is evidence that the movement of water across the root, can have an effect on the delivery of ions into the stele. This has been observed in both exuding root systems and in intact plants where the water movement driven by transpiration [22]. The transport of ions across the root is depend ent upon active ion uptake across the cortical plasma membrane at some point followed by transport through the symplasm, at least across the endodermis. Water must also follow a similar route although the relatively low resistance of the cell wall pathway favours water movement and thus both water and ions will move together through the endodermal cytoplasm Any increase in water flux might therefore increase the rate of ion flux across the root. This has been observed in experiments where the slowing down of the water flux with mannitol results in a decreased K^+ transport

28

across the root, although a portion of the K^+ flux remained independent of the rate of water movement [12]. The rate of water movement may also increase the removal of ions from the lower ends of the xylem conduits, resulting in a steeper gradient for the release of ions within the stele.

It has been suggested that the uptake of ions by the root, although undoubtedly active at low external ion concentrations, may become 'passive' at higher external levels. This possibility requires that the plasma membrane becomes permeable at high external ion levels and thus no longer presents a high energy barrier for ion movement. This unlikely situation has been clearly demonstrated not to occur and ion uptake by plant roots is an active process over a wide range of external concentrations, although kinetic analyses indicate that different uptake systems may be involved [35].

In simple terms the movement of water across an exuding root system may be calculated as

$$\phi_v = L_p \left(-\sigma RT \Delta c_j \right) \tag{3.1}$$

where ϕ_v is the volume (water) flux into the root, $m\ s^{-1}$, L_p the hydraulic conductivity of the root, $m\ s^{-1}\ bar^{-1}$, σ the reflection coefficient of the osmotic barrier within the root, R and T as before, and c_j the solute concentration difference the xylem exudate and the external medium. An additional non-osmotic term has been sometimes added to the above equation to account for water flow in the apparent absence of a water potential gradient [63], although this anomaly may merely reflect that the above equation oversimplifies the exudation process. A model invoking a standing osmotic gradient has been proposed which partially resolves the above anomaly [6].

When the water movement across the root is driven by transpiration an hydrostatic pressure term ΔP must be added to the total driving force in the water to give

$$\phi_v = L_p \left(\Delta P - \sigma RT \Delta c_j \right). \tag{3.2}$$

This equation is also difficult to apply as L_p apparently increases with increased values of ΔP [83].

The ion flux emerging from an exuding root system ϕ_j is simply the product of the exudation volume flux ϕ_v and the concentration of the ion in the exudate, c_j

$$\phi_j = \phi_v c_j \tag{3.3}$$

ϕ_j is expressed in units of $mol\ m^{-2}\ s^{-1}$, where areas can be calculated, or $mol\ gFW^{-1}\ s^{-1}$ for highly branched root systems.

There is evidence that both ϕ_v and ϕ_j can be modified by plant growth substances such as kinetin and abscisic acid [28] (see Table 3.1). The mode of action of these hormones is not well understood at present but it would appear that they have a marked effect on membrane permeability and on transport through the symplasm. In this respect it is noteworthy that ϕ_{Ca} is the least affected by hormone treatment and Ca^{2+} is poorly transported through the symplasm. The results with hormones also give a clue to

29

Table 3.1 The effect of hormone treatment on the volume flux, ϕ_v^{oi}, and ion flux, ϕ_j^{oi}, from 3-day old maize roots, 24 h after excision [28]

	ϕ_v^{oi}	ϕ_K^{oi}	ϕ_{Ca}^{oi}	ϕ_{Cl}^{oi}
	(m^3 m^{-2} s^{-1})	(nmol m^{-2} s^{-1})		
Control	5.0	110.5	9.2	115.5
Abscisic acid	7.4	151.7	8.1	127.3
Kinetin	1.0	16.1	2.9	15.4

possible integration processes within the intact plant which are not normally manifested in the isolated exuding root system. The hydraulic conductivity of onion roots is doubled by the presence of a shoot, implicating hormone mediation in a root/shoot interaction [58].

3.3 Transfer cells

It has long been known that plant cell walls can form intrusions or ingrowths into the cytoplasm of certain cells. These inwardly directed projections of the cell wall create invaginations and infoldings of the plasma membrane thus increasing the surface area for solute transport. Such cells have been termed transfer cells by Gunning and Pate, [48] who have made an extensive survey of their distribution in terrestrial plants. Gland cells, epidermal cells, phloem parenchyma, companion cells, xylem parenchyma, pericycle cells, may all produce wall ingrowths and an amplified plasma membrane; in such conditions they may be termed transfer cells. The wall ingrowths are a specialized form of secondary wall, are not penetrated by the middle lamella and the plasma membrane is in intimate contact with these ingrowths to form a 'wall-membrane apparatus'. Mitochondria are abundant in such cells as are cisternae of the ER often lying close to the involuted plasma membrane.

Transfer cells are often polarized, the wall ingrowths being restricted to these walls abutting conducting elements of the xylem and phloem. There is frequently an association between transfer cells and these long-distance transporting systems (Fig. 3.6). Xylem transfer cells are the most abundant, existing in the majority of plants so far investigated. They are modified xylem parenchyma cells with ingrowth only on those walls adjacent to tracheary elements, found mainly at nodes and restricted to foliar traces where gaps are created in the vascular cylinder. Nodes which have xylem transfer cells commonly possess phloem transfer cells although generally the latter are best developed in those parts of the vascular network devoid of xylem transfer cells. It has been postulated that the siting of these xylem transfer cells facilitates transport of solutes out of the transpiration stream into the symplast of the tissues and organs adjacent to a departing leaf trace. Such transfer cells have been observed to remove nitrogenous materials in a selective manner from ascending xylem sap.

Phloem transfer cells are found in the minor veins of the leaves of some species. They are generally of two types, designated A and B. The A-type is most common and is specifically associated with sieve elements. It has a

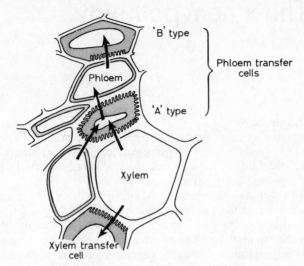

Fig. 3.6 'A'- and 'B'-type phloem transfer cells and a xylem transfer cell with the major postulated transfers indicated by arrows [48].

dense cytoplasm, an even distribution of wall outgrowths except where plasmodesmata connect it with the adjacent sieve element. The B-type is found widely scattered throughout a vein. Wall outgrowths are restricted to those walls oriented towards sieve elements or A-type cells. It is possible that both of these transfer cell types may play an important role in the loading of assimilates into the sieve elements, particularly if such loading was from the free space. Transfer cells are rare within root vascular tissue although they occur in the root nodules formed when *Rhizobium* infects a leguminous plant, and presumably facilitate the movement of nitrogenous materials to the host plant.

Bibliography
Bowling, D.J.F. (1976), *Uptake of Ions by Plant Roots*, Chapman and Hall, London.
An outline of the direct knowledge we have about ion uptake by roots.
Epstein, E. (1972), *Mineral Nutrition of Plants, Principles and Perspectives*, Wiley, New York.
A personalized account of plant mineral nutrition by a leading researcher in the field of ion uptake.

4 The xylem pathway

4.1 Xylem structure

For transport over relatively short distances the cell to cell pathway is adequate but, in higher plants in particular, long-distance transport of water and dissolved solutes is through the specialized conduits of the xylem and of the phloem (see p. 40).

Within xylem the two types of conduits or tracheary elements found are composed of tracheids or of vessel elements. Both of these cell types have thick, lignified secondary walls and lose their cytoplasm during differentiation, resulting in dead, pipe-like members which permit the movement of relatively large volumes of solution with no cytoplasmic or membrane resistance to overcome. Whereas the more highly evolved angiosperms contain both tracheids and vessels, gymnosperms and lower vascular plants have only tracheids. Thus we may consider tracheids to be more primitive than vessels and this is reflected in the relative efficiency of these elements in the conduction of water and solutes (see p. 64).

Tracheids are narrow ($< 50 \mu m$) lignified cells ranging in length from $100 \mu m$ to 10 mm or more, with abundant pits and perforated end walls which are connected through pit pairs with adjacent elements. The pathway for long-distance transport through tracheids is thus somewhat tortuous in comparison with the straight through route provided by the more evolutionary advanced vessel elements described below.

The vessel elements are generally shorter than tracheids and usually thicker, having a diameter ranging from $10 \mu m$ to as much as $700 \mu m$. Individual vessel members often lose their sloping end-walls completely so that a column of vessel elements will form a hollow tube, the xylem vessel, thus providing a low resistance pathway through which water and solutes can move more rapidly than through tracheids (Fig. 4.1).

In addition to vessels and tracheids the xylem is composed of wood fibres and xylem parenchyma. The fibres are long, thin, heavily lignified cells which also usually lose their cytoplasm during differentiation. They contribute to the structural support of a plant and are not directly involved in water and solute transport. The living parenchyma of the xylem are important in that they permit lateral movement of water and solutes into and out of the conducting cells. Their other major function is the storage of carbohydrates, mainly as starch, which accumulates in these cells during one growing season and are utilized during renewed cambial activity at the start of the next season. In addition, xylem transfer cells (see p. 30) are found in the petioles and stems of herbaceous dicotyledons, but are uncommon in leaf veins.

The functional life of a vessel or tracheid as a pathway for water and

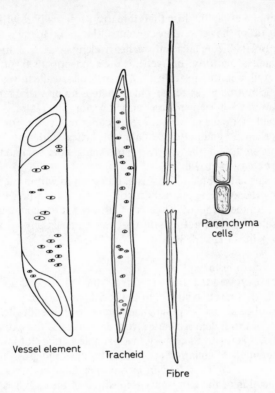

Vessel element   Tracheid

Parenchyma
cells

Fibre

Fig. 4.1 Representative cell types from xylem.

solute transport may be several years but eventually the older xylem
elements cease to translocate, the lumen often becoming blocked by the
production of resins and gums. In many species it is only the most recently
formed xylem which is involved in transport, the older parts of the tissue
playing a structural role only.

As indicated above, the differentiation of tracheary elements is directed
towards producing a non-living conduit. When a potential tracheary element
has been formed by division from the meristem, the vacuoles enlarge and
coalesce. At the same time rapid and extensive growth occurs, resulting in
the restriction of the organelles to the ends and central part of the elongat-
ing cell. The major differentiation of the cells is in the formation and
lignification of secondary walls. The endoplasmic reticulum (ER) and
Golgi apparatus play an important role and microtubules are abundant,
strategically placed for the guidance of the vesicular derivatives of these
organelles to the cell surface. Evagination of the plasma membrane may
also contribute materials for wall formation. There is some evidence that
the ER only allows wall formation to occur where it does not directly shield
the plasma membrane, thus providing a template for secondary wall deposition. Before completion of this growth and secondary wall formation the
cytoplasm begins to break down. Mitochondria and plastids degenerate and

33

the cell membranes rapidly lose their integrity, eventually resulting in the complete death of the cell and its contents. The resulting hollow tube is not completely isolated from adjacent tracheary elements, the lumina being interconnected at maturity. In tracheids the connection is through remnants of primary wall modified to form a 'pit membrane', while in vessels the connection is by true pores across which there is no pit membrane. Areas of such pores in vessel elements are termed perforation plates.

The secondary thickening is not evenly deposited on the primary wall of a tracheary element, and patterns of annular, helical, reticulate or pitted arrangements are found, the first two types being more common in tissues undergoing extension growth.

There is a paucity of plasmodesmatal connections within the tracheary elements and the xylem at maturity is completely apoplastic with no direct continuity with the symplast. Thus as a pathway for the conduction of water and solutes it does not present the same complex physiology as does the phloem pathway (see p. 39).

4.2 Ion movement in the xylem

Once ions have crossed the root and entered the xylem conduits they are carried in the transpiration stream to the aerial parts of the plant where they are absorbed across the plasma membranes of the leaf cells. The composition of this xylem sap varies with the nature of the external medium but in general it is found to be a relatively simple solution of inorganic ions, with small amounts of amino acids, carboxylic acids and growth substances. Root pressure exudate from *Ricinus* grown on a dilute culture solution gives an indication of the inorganic composition of the xylem sap (Table 4.1). It can be seen that there is an excess of cations over anions. Presumably the balance is provided by organic ions.

In *Ricinus*, NO_3^- is the major form of nitrogen (N) within the xylem, but in many plants nitrate reductase activity in the roots results in a high level of amino acids in the xylem sap. In apple, for example, virtually all the transported N is organic, with aspartate, asparagine, glutamine, glutamate, other amino acids and peptide-like substances accounting for most of the N present [17]. The major endogenous growth regulators, auxin, gibberellins, cytokinins and abscisic acid (ABA) have been identified in xylem saps, and it is possible their presence may reflect some form of control system operating between root and shoot (see p. 30).

There is some evidence that there may be active removal of certain ions, particularly Na^+, as the transpiration stream moves through the xylem of the root. In maize Na^+ levels of xylem exudate are considerably lower when collected from longer lengths of root, whereas the level of Na^+ in the root tissue increases with increasing length [108]. EPM studies have indicated that Na^+ is accumulated in modified xylem parenchyma cells [132] which may be responsible for the regulation of the ion content of the transpiration stream as it leaves the root. Na^+ removal also occurs in stem tissues [68] contributing to the low $[Na^+]$ in the leaves of many plant species. Lateral movement of ions out of the xylem presumably takes place throughout the passage from the absorbing roots to the leaves, such

Table 4.1 Concentrations of the major ions in the external solution and root pressure exudate from *Ricinus*, together with calculated Nernst potentials and driving forces [24]

Ion	c_j^0 (mol m^{-3})	c_j^i (mol m^{-3})	E_{N_j} (mV)	ΔE_j (J mol^{-1})
K$^+$	0.60	5.6	-56	$+400$
Na$^+$	0.33	0.74	-20	-3200
Ca^{2+}	0.84	10.1	-31	-2100
Mg^{2+}	0.45	3.9	-27	-2500
NO$_3^-$	1.45	12.0	$+53$	$-10\ 500$
Cl$^-$	0.14	1.03	$+50$	$-10\ 200$
SO$_4^{2-}$	0.69	2.00	$+13$	-6400
H$_2$PO$_4^-$	0.11	1.38	$+63$	$-11\ 600$
HPO$_4^{2-}$	0.13	0.56	$+18$	-6900

Mean value $E_M = -52$ mV

Table 4.2 Comparison of [K$^+$]/[Na$^+$] in shoots of whole barley seedlings and exudate from de-topped seedlings [97]

Concentration (mol m^{-3})		Ratio of [K$^+$]/[Na$^+$]			
K$^+$	Na$^+$	Solution	Roots	Shoots	Exudate
0.5	9.5	0.052	0.8	3.4	2.1
1.0	9.0	0.11	1.1	4.0	3.1
2.5	7.5	0.33	1.9	8.1	7.1
6.0	4.0	1.5	6.7	18	22
8.0	2.0	4.0	8.9	30	36

ions often appearing in the phloem tissues. In trees considerable distances are involved and the xylem sap composition may be markedly altered before it reaches the leaves, in many cases transfer cells (see p. 30) being involved in a selective extraction of ions [48]. The walls of the xylem conduits, in common with other cell walls, contain numerous fixed negative charges which can retain cations. Divalent cations are preferentially bound and their movement through the xylem is thus retarded. The movement of Ca^{2+} may be envisaged as a continuous displacement as in an ion-exchange column [14]. When Ca^{2+} is supplied to plants in a chelated form (as calcium ethylenediaminetetraacetic acid, Ca EDTA), although the initial uptake is slower, the Ca^{2+} absorbed in this form is more mobile as it is not adsorbed on the exchange sites in the xylem [67].

As a result of selective accumulation in the initial uptake, followed by the removal of ions in the xylem, the xylem sap which ultimately reaches the mesophyll cells through the apoplast in the leaf has a different composition to that experienced by the root cortical cells. The level of K$^+$ may be considerably higher than that of the external solution, while Na$^+$ may be reduced to a very low level. This is reflected in the high ratio of [K$^+$]/[Na$^+$] in the shoots of many plants (Table 4.2).

4.3 Regulation of leaf nutrient content

The leaf may be considered as a closed system receiving solutes from the xylem and, in early stages of development, from the phloem. Loss from the system will be by export through the phloem or by processes such as guttation, leaching, and, in certain plants, by excretory salt glands. Thus regulation of ion content in the leaf involves many processes as indicated in Fig. 4.2. The observed ion content can change in response to variation of input through xylem or phloem, export in the phloem, or loss across the cuticle.

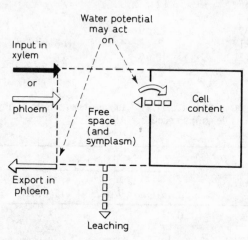

Fig. 4.2 Regulation of ion content in the leaf [97].

Uptake of ions by the leaf mesophyl cells has received less attention than uptake by the root. However, it appears that the K^+ uptake kinetics of leaf tissue are similar to those found in root tissue, are insensitive to light, and therefore coupled to energy metabolism independently of photosynthesis [112], but as in algae (see p. 16) some light-stimulated ion fluxes have been found [70].

As ions are continually supplied to the leaf in the transpiration stream they often receive considerably greater levels of certain ions than they require for growth and metabolism. Some excess ions are accumulated into the vacuoles of the leaf mesophyll cells where they serve an osmotic function, but in many plants the surplus is exported from the leaf in the phloem (see p. 39). This transfer of ions from xylem to phloem within the leaf may involve transfer cells (see p. 30). Ions such as Ca^{2+}, which are relatively immobile in the phloem, continue to accumulate within the leaves, often in the form of precipitates, crystals of calcium oxalate and calcium carbonate commonly occurring in the vacuoles and cell walls of many species.

Leaching of solutes from leaves by rain or mist removes considerable amounts of certain ions (Table 4.3). In general, cations are leached preferentially to anions and the leachate is slightly alkaline [121].

Table 4.3 Effect of rain in leaching nutrients from leaves of deciduous forest trees [97]

| Element | Quercus petraea | | Maple/birch/beech | |
	Content of leaves at fall (μmol gDW^{-1})	% Removed by rain during season	Content of leaves, standing crop (μmol gDW^{-1})	% Removed by rain during season
N	760	−1.8	1700	13
P	19	40	19	12
K	69	240	270	91
Na	19	1200	0.8	500
Ca	155	42	160	30
Mg	38	120	46	50
S	–	–	15	370

Leaching has its converse in the foliar feeding of nutrients widely practised in agriculture. Under humid conditions a positive root pressure may develop which results in droplets of xylem sap being extruded from the leaf, the process of guttation. This guttation fluid, usually a dilute solution of inorganic ions and organic compounds, will contribute to the overall loss of ions from the leaf, any subsequent rainfall leaching these solutes from the system.

Loss of solutes by leaching and guttation are somewhat fortuitous processes, depending upon rainfall and humidity, over which the plant has no control. However, certain halophytic plants have evolved a mechanism for the excretion of NaCl from specialised salt glands, enabling the plants to withstand the high salinity of their environment. In the mangrove *Agialitis annulata* the excretory process is very specific, the excreted fluid containing 450 mol m^{-3} Cl$^-$, 355 mol m^{-3} Na$^+$ and only 27 mol m^{-3} K$^+$. Thus K$^+$ is effectively discriminated against by the gland giving a lower [Na$^+$]/[K$^+$] ratio in the leaf [7], as shown by the following [Na$^+$]/[K$^+$] ratios

excreted solution	13:1
leaf tissue	3:1
xylem exudate	8:1

A simpler salt excreting system is the salt hairs found on the leaves of many chenopod species. These hairs are usually differentiated into an excretory stalk cell and a salt-accumulating bladder cell. Elimination of salt from the interior of the leaves into the vacuoles of these epidermal bladder cells can reduce the salt level in the photosynthesising tissues. Salt glands are more complex, comprising between 2 to 40 cells or more. Collecting cells may be distinguished from the salt-eliminating excretory cells.

The physiological mechanism of salt excretion is probably similar in both glands and hairs (Fig. 4.3). A symplasmic continuum exists between the gland cells and the surrounding mesophyll cells through which ions are

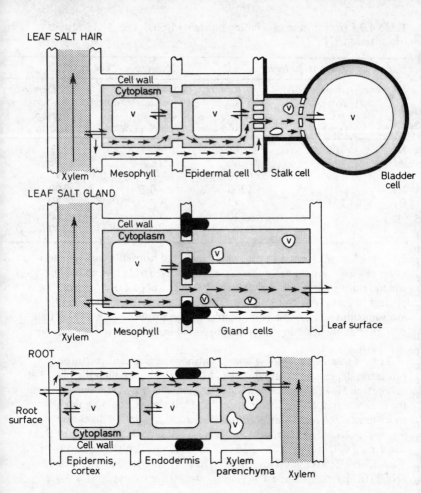

Fig. 4.3 Simplified structural model of a leaf salt hair, of a leaf salt gland, and of a root, showing certain analogies with respect to the long-distance transport of solutes [75].

transported, cuticular envelopes blocking the apoplast pathway. Salt excretion by the glands is a metabolically dependent process but the site of the active ion pumps involved has not been resolved. Excretion could be driven by ions pumped into the symplast within the leaf or out of the symplast within the gland cells [75].

Bibliography

Baker, D.A. and Hall, J.L. (1975), *Ion Transport in Plant Cells and Tissues*, North Holland Publishing Co. Amsterdam.
A comprehensive, advanced multi-author text on all aspects of ion transport in plants.
Lüttge, U. and Pitman, M.G. (eds) (1976) *Encyclopedia of Plant Physiology*,

N.S. Volume 2B. Springer-Verlag, Berlin, Heidelberg, New York.
An excellent multi-author treatise on ion transport at the tissue, organ and whole plant level.

5 The phloem pathway

5.1 Experiments to determine the pathway of assimilate translocation

When the sieve element was discovered in the bark of woody plants a transporting function was proposed for these chains of long pipe-like cells [53]. Exudation of sap from cuts made in the bark of trees was found to contain up to 33% sugars and also small amounts of minerals and nitrogenous substances. The role of the phloem as a conducting tissue was apparently established.

However Sachs proposed that only nitrogenous materials moved through the sieve elements and that sugars moved by diffusion in the phloem parenchyma. Sachs' influence was so strong that his views prevailed throughout the nineteenth century despite a clear demonstration that the starch-containing parenchymatous layer that surrounds the phloem of many species does not conduct carbohydrates [59].

By the turn of the century improvements in histological techniques had resulted in descriptions of sieve element development in both gymnosperms and angiosperms. Work with a leaf-mining moth larva demonstrated that whenever the larva had eaten its way through the phloem carbohydrate movement out of that leaf was interrupted. However, the pathway controversy was compounded by the calculation that during its development a potato tuber would need sugars to move through the stolon at 0.4 m h^{-1} or more (see p. 55), and the conclusion that sieve elements were unsuited to the transport of substances at such speeds [33]. Transport of carbohydrates was proposed to be the result of a reverse movement through the xylem!

Elegant experiments by Mason, Maskell and Phillis, in the late twenties finally resolved the pathway problem. They found that both nitrogenous materials and carbohydrates were transported through the innermost layer of the bark which contains the sieve elements [82].

It is to the credit of these early investigators that their pioneer studies, made without the availability of radioisotopes, have been subsequently verified. The use of isotopes has greatly facilitated the study of transport pathways, demonstrating upward movement of ions predominantly through the xylem (p. 32) and the distribution of assimilates through the phloem. Such experiments often involved surgical separation of the transporting tissues which may cause damage and alter the normal pattern of translocate movement.

This problem has been partially overcome by feeding $^{14}CO_2$ to plants and utilizing certain phloem-feeding aphids [127]. These insects insert their stylets into specific plants, avoid all other cells until a sieve element is encountered, puncture it and then feed passively on the phloem sap. As the aphids enjoy a large surplus of sugars they release honeydew which can be analysed for radioactivity. If label is present then the translocate has moved to that particular region of the plant, or has been induced to move by the aphid forming a sink. If pure phloem sap is required the aphid is anaesthetia with a stream of CO_2 and excised from the inserted stylet. (The anaesthetic is to prevent the insect from moving during this operation rather than for altruistic pain-relieving purposes!) Pure phloem sap then flows from the stylet stump at up to 5 μl h^{-1}, the tapped sieve element being refilled 10 times each second. These observations provide strong evidence that the sieve element is the normal pathway for translocated assimilates.

Microautoradiography has provided a powerful technique for translocati studies. Radioactive tracers are fed to the plant and, after allowing time for transport, thin frozen sections are cut, a thin layer of photographic emulsio placed over the tissue slice on a glass slide, and the 'sandwich' left in the darkness to allow image formation. The film is then developed while still in contact with the tissue, and microscopic examination permits the radio-activity to be observed as an image superimposed upon the section. This technique has limitations, and problems arise when soluble compounds such as sugars are being investigated, some diffusion taking place even in frozen material. It is also difficult to resolve the exact location of ^{14}C within cells as small as sieve elements and interpretation of microautoradiographs requires experience.

Evidence that sieve elements are the channels of translocation is largely indirect. Perhaps the only direct evidence is the observation that eosin, which induces callose formation thus blocking sieve plate pores (p. 43), causes a concomitant cessation of translocation [107]. Consideration of the cell types present in the phloem also indicates that sieve elements provide the translocation pathway. Phloem fibres are not always present, companio cells also are occasionally absent and are often discontinuous from one end of a sieve element to the other. Thus sieve elements present the logical choice with their highly specialized structure when compared with the largely undifferentiated phloem parenchyma.

Further indirect evidence of the major role of sieve elements is provided by the parallel evolution in the conducting tissues of giant kelps. These large brown algae also have an extensive translocation requirement; assimilates from well-illuminated fronds near the surface must be transporte to poorly-illuminated parts at lower depths. ^{14}C-labelled assimilates move at up to 0.78 m h^{-1} through longitudinally directed sieve filaments, which are morphologically similar to sieve elements [90].

5.2 Structural design of the sieve element
Sieve tubes occur only in angiosperms; the phloem of gymnosperms and pteridophytes contains less specialized sieve cells with a similar function. These sieve cells are not vertically aligned and the sieve areas are located in

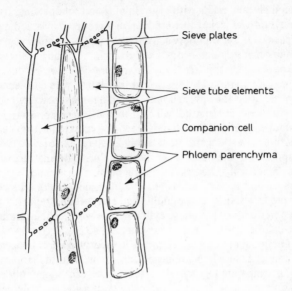

Fig. 5.1 Phloem in longitudinal section.

lateral walls and are rarely found in end walls. The sieve tube is composed of files of sieve elements which are highly specialized cells with a complex differentiation and development (see below). The phloem is composed of sieve elements, which occupy some 20% of the tissue (in some trees 60–70%) the remainder being companion cells, parenchyma and fibres (Fig. 5.1). Mature sieve elements are typically 100–500 μm in length with a diameter of 20–30 μm, characterized by having perforated end walls (the sieve plates), a degeneration of the normal cellular organelles, and a unique proteinaceous content termed P-protein (phloem protein).

Sieve tubes which often extend from leaf to root, develop rapidly by the almost simultaneous differentiation of a longitudinal row of cambial cells. Many sieve tubes have a row of associated companion cells, formed by a final, unequal longitudinal division of each cambial initial before development of the sieve tube begins. Sometimes sieve tubes are associated with two or more rows of companion cells whereas in a few species companion cells are absent.

At first the sieve element cells appear to be normal elongated parenchyma, with streaming cytoplasm, a prominent nucleus and a clearly distinguishable vacuole bounded by a tonoplast. Abundant mitochondria, plastids, ribosomes, dictyosomes and a well-developed ER are present. The cells will accumulate dyes such as neutral red and can be plasmolysed.

An early sign of differentiation is the appearance of prominent plasmodesmata, especially in those regions of the wall destined to become sieve plates. Sieve pores develop by the gradual breakdown of wall materials around groups of plasmodesmata while at the same time callose is deposited on the developing sieve plate. Callose, a relatively impermeable β–1, 3 linked

41

unbranched glucan, reacts with resorcin blue and other stains. It appears first as a cylindrical jacket surrounding each group of plasmodesmata and later spreads over the wall between the sieve pores. At this stage the nucleus begins to degenerate, dictyosomes and mitochondria become less prominent and the tonoplast disappears. Thus the mature sieve element loses not only nuclear information but also ribosomal translation sites and secretory dictyosomes. Strands of cytoplasmic material remain scattered sometimes appearing to connect longitudinally through the sieve pores. The membrane-bound tubules (transcellular strands) reported to extend from cell to cell in the mature sieve elements are possibly artefacts [116].

Sieve pores become blocked with callose during dormant periods, reactivation normally occurring with the break of dormancy when callose is redissolved and sieve elements become functional again. Callose formation can be rapid in response to wounding and early studies reported that functioning sieve pores contained callose cylinders. However, when suitable precautions are taken callose is not found in any quantity [3].

The functional life of sieve elements in dicotyledons is seldom more than a year, vascular tissues being renewed continuously in the tropics and annually in temperate regions. However, arborescent monocotyledons such as palms do not renew their stem vascular tissues and sieve elements in the lower portions of these may have a functional life of over 50 years.

Translocation through the sieve elements occurs only as long as they are living. If the cells in a short section of stem are killed by a jet of steam or by other means, phloem transport ceases, although transport in the xylem system usually proceeds without interruption. Presumably the companion cells must also be living for translocation to occur, as they are assumed to be essential for the maintenance of life in the sieve elements and may play a role in translocation itself, but at present little is known of their specific function. In gymnosperms the sieve cells have associated albuminous cells not derived from the same cambial mother cell as the sieve cell, and therefore not strictly analogous with companion cells, probably performing a similar function.

A major postulated role of the companion cell is the maintenance of sieve element structures and osmotic balance. As sieve elements have neither nucleus nor ribosomes, protein synthesis is dependent on the companion cell. The interdependence of these two cells is reflected in the fact that the companion cell dies with the sieve element whereas adjacent parenchyma continue living. Sieve elements have numerous plasmodesmatal connections with their companion cells. On the sieve element side the pore is lined with callose while on the companion cell side numerous branches occur which anastamose in the mid-line of the wall. Such branched plasmodesmata, characteristic of connections between companion cells and sieve elements, probably permit a massive transfer of assimilates. In leaves, companion cells are often larger in diameter than sieve elements, again probably reflecting the important function of these cells in the loading of sieve elements. In the stem, however, the diameter of sieve elements may be up to 100 times greater than that of companion cells.

Another unique feature of sieve elements is their unusually thick walls,

hich present a glistening appearance when viewed under the light micro-
cope, termed nacreous thickening, which results from the high degree of
ydration of their constituents. These walls sometimes occupy an appreci-
ble fraction of sieve elements, which led some early investigators to suggest
ansport through the wall itself. Certainly the high degree of hydration
dicates a wide spacing of the cellulose polymers which could result in a
all structure capable of accommodating rapid solution flow, and recently
is possibility has been reconsidered (see Section 6.2). However, in grapes
e sieve element walls are relatively thin and do not have a nacreous
ppearance, making it unlikely that the walls provide a significant pathway.

2.1 The sieve plate: open or closed pores?

he feature which gives sieve elements their name is the perforated end
all of each element, the development of which has been outlined above.
he sieve plate is normally about 1 μm thick, the pores have a diameter of
.1−5 μm depending on the species and occupy some 50% of the end wall
rea. The sieve plate is presumed to have a function directly related to
ranslocation since they are numerous in the species more advanced in
volutionary terms.

When plants are grazed by herbivores or otherwise damaged, the sieve
lement contents would leak, particularly as they are subjected to relatively
igh positive pressures, but such leakage rarely occurs. The sieve plate pores
apidly block with callose and P-protein, sealing the system and preventing
ny excessive loss of contents. Some workers have proposed that sieve
lates may provide a framework to subtend transcellular strands or, alter-
atively, to provide a support for a charged matrix of P-protein (see
ection 6.2).

Central in the controversies regarding sieve element structure is the
uestion of contents within the pores of functioning sieve plates, some
nvestigators advocating 'open' pores devoid of any contents while others
rgue that the pores are 'closed' during translocation. The resolution of
his problem is critical to an evalution of proposed mechanisms of trans-
ocation (p. 66) some theories requiring 'open' and others 'closed' pores.

This problem remains unresolved, the sieve element being easily
amaged and its contents displaced during preparation for structural
tudies. Pressure release upon sectioning may blow the contents into (or
ut of) the pores, completely reversing the appearance from the function-
ng condition. Different fixatives have produced contradictory EM pictures
f pore contents. When sieve elements are fixed callose rapidly forms and
nay block the sieve pores or change their shape. Thus it is not yet possible
o accurately describe the condition of pores in functioning sieve elements
ith any great confidence, although many try.

The problem is even further complicated by another difficulty; the
tages of their differentiation when sieve elements translocate most
ffectively is not known. Perforation of the sieve plates is assumed to
recede translocation, but clear evidence for this is unavailable. As yet
here is no method for showing that a sieve element seen in an electron
nicrograph was translocating when in the living state.

5.2.2 *P-protein*

EM studies of sieve element structure reveal that the substance originally described from optical microscopy as 'slime', consists of a proteinaceous material termed P-protein. Plant slimes are usually carbohydrate and the term slime is misleading. The most commonly encountered form of P-protein is tubular, in some species tubule formation being preceded by the formation of a fibrous P-protein. In mature sieve elements the P-protein is usually in the fibrillar form, giving a characteristic beaded appearance. In some legumes the P-protein is crystalline, in an intermediate state between tubular and dispersed beaded fibrils.

P-protein is formed within a special cell structure, the P-protein body, which disaggregates at the time when the nucleus and tonoplast become disorganized resulting in the dispersal of filaments throughout the sieve element lumen. Tubular elements ranging from 18–23 nm diameter are frequently found and are designated P1-protein. P1-protein tubules are often closely associated with stacks of membranes, are linked by short cross-bridges, and occasionally display an hexagonal packing. After dispersal they are in the finer (6–12 nm diameter), filamentous and striated P2-protein form [32]. It has been suggested that the P-protein filaments consist of either one or two helically arranged chains of globular subunits, combination and separation of these chains, accompanied by stretching and loosening of the helices, leading to the different structures observed [91].

EM studies reveal P-protein in the lumina of sieve elements and in the pores of sieve plates. Most of the structures reported earlier using the optical microscope were also probably P-protein. Whether the filaments are attached or free to move is not known. If free to move it might be expected that a higher accumulation of P-protein would occur in the 'sink' end of the sieve tubes but this is not observed. Certainly when there is a pressure release, as occurs when material is sectioned or injured in some way, P-protein accumulates on one side of the sieve plate and in the pores, in which position it presumably aids the blockage and sealing of damaged sieve elements. If the P-protein is anchored then such a pressure release must break the attachment.

P-protein may provide a structural arrangement facilitating translocation or may function purely as a sealing mechanism. It has been suggested that P-protein may be directly involved in the production of a motive force for translocation (see p. 70). Whatever its role within the sieve element this function must be restricted to the angiosperms, since P-protein is absent from the sieve cells of gymnosperms [131].

5.2.3 *Membranes and organelles*

During differentiation of the sieve element the nucleus disintegrates, although the significance of this is obscure. Nucleoli can still sometimes be found after nuclear breakdown. Stacks of ER cisternae of various sizes become associated with the nuclear envelope prior to nuclear disintegration after which these stacks continue to enlarge and sometimes form large aggregates. These aggregates eventually form loose patches which combine with individual cisternae to form a peripheral network of cisternae just

44

inside the plasma membrane and closely associated with it, termed the sieve element reticulum. The plasma membrane retains its selective permeability during this differentiation as demonstrated by the observation that sieve elements will plasmolyse.

Mitochondria, often found anchored in the parietal layer of mature sieve elements, remain unmodified during the differentiation of the cell. Reported degenerative changes in the mitochondria during differentiation appear to be artefacts [31]. Their presence in mature sieve elements has led to the suggestion that mitochondria play a vital and regulatory role in assimilate transport, but such a role will remain obscure until the translocation mechanism has been elucidated. Companion cells contain large numbers of mitochondria and probably the sieve elements are combined into functional units with these cells (p. 42).

Starch-containing plastids are often found in the parietal layer of sieve elements. In dicotyledons the plastids normally contain a highly branched form of starch with numerous $\alpha-1, 6$ linkages which gives a specific reddish-brown staining reaction with iodine. Sieve element plastids show little development of their internal membrane system and although some enlargement of the plastids occurs during differentiation this membrane system remains poorly developed. The plastids release their starch grains when damaged and these grains, whose function is not known, are often found blocking the sieve plate pores.

At the time when the P-protein bodies disaggregate (see below) the tonoplast changes its permeability properties and starts to disintegrate, resulting in a greater hydration of the cytoplasm which then appears less dense in EM preparations. The continued presence of a tonoplast in mature sieve elements lining each pore of the sieve plate has been reported [115] but the tonoplast is normally absent. The term vacuole is not used, the central region of the sieve element being described as a lumen.

During the development of the sieve element the P-protein aggregates to form P-protein bodies, the form of which varies in different plant species. These P-protein bodies increase in size during development of the sieve element by the addition of more P-protein. In legumes, bodies of 30 μm with long sinuous 'tails' have been observed, and were described by earlier workers as 'flagellar inclusions'. P-protein bodies are membraneless but their integrity is maintained until the dispersion stage, which occurs at the same time as the breakdown of the nucleus, tonoplast and reorganization of the ER (p. 42).

5.2.4 Current interpretation of the structure of the functioning sieve element

The structure of the functioning sieve element is unresolved at present and on the basis of present evidence at least six different arrangements are possible (Fig. 5.2).

Fig. 5.2a represents a sieve element in which the parietal cytoplasm covers the sieve plate pores so that each sieve element lumina is a discrete entity. Fig. 5.2b depicts the 'open' pore situation in which the lumina are continuous through the sieve plate pores for a file of elements. Fig. 5.2c

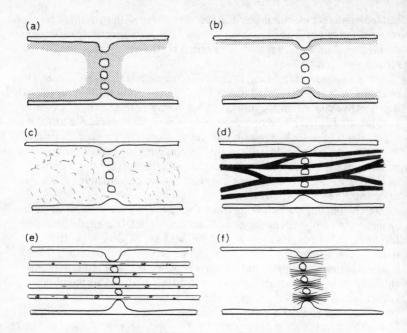

Fig. 5.2 Possible arrangements of the contents of translocating sieve elements [126].

represents a network of P-protein dispersed uniformly throughout the lumina of the sieve elements and continuous through the sieve plate pores. In Fig. 5.2d aggregated strands of P-protein are shown running through the pores of the sieve plate, perhaps with some associated membranes of the endoplasmic reticulum. A similar pattern is shown in Fig. 5.2e where proposed membrane trancellular strands [116] are represented. The organized association of P-protein within the sieve plate pores [114] is shown in Fig. 5.2f. Some of these proposed structures are not mutually exclusive and combinations of these patterns may be envisaged. However, improved EM techniques will hopefully soon provide the elusive structure of the functioning sieve element. This may already have been obtained but an unequivocal model is required that will end the current controversy. This difficult of obtaining an exact picture of the internal structure of functioning phloem has added greatly to the problems of resolving the mechanism of translocation.

5.3 Composition of phloem sap
Analysis is usually made on phloem saps, which can be collected from a number of plant species by various methods. These involve making incisions in the bark of trees, cutting the stems of some cucurbits and *Ricinus,* the inflorescence stalk of *Yucca* and the fruits of some legumes. Alternatively the aphid stylet technique has been used for collecting sieve-tube sap (p. 40).

Table 5.1 The composition of phloem exudate obtained from *Ricinus* plants. Concentrations are expressed in mg ml^{-1} and also in eq m^{-3} or mol m^{-3} where relevant [51]

	mg ml^{-1}	
Dry matter	100–125	
Sucrose	80–106	
Reducing sugars	Absent	
Protein	1.45–2.20	
Amino acids	5.2 (as glutamic acid)	35.2 mol m^{-3}
Keto acids	2.0–3.2 (as malic acid)	30–47 eq m^{-3}
Phosphate	0.35–0.55	7.4–11.4 eq m^{-3}
Sulphate	0.024–0.048	0.5–1.0 eq m^{-3}
Chloride	0.355–0.675	10–19 eq m^{-3}
Nitrate	Absent	
Bicarbonate	0.010	1.7 eq m^{-3}
Potassium	2.3–4.4	60–112 eq m^{-3}
Sodium	0.046–0.276	2–12 eq m^{-3}
Calcium	0.020–0.092	1.0–4.6 eq m^{-3}
Magnesium	0.109–0.122	9–10 eq m^{-3}
Ammonium	0.029	1.6 eq m^{-3}
ATP	0.24–0.36	0.40–0.60 mol m^{-3}
Auxin	10.5×10^{-6}	0.60×10^{-4} mol m^{-3}
Gibberellin	2.3×10^{-6}	0.67×10^{-5} mol m^{-3}
Cytokinin	10.8×10^{-6}	0.52×10^{-4} mol m^{-3}
Abscisic acid	105.7×10^{-6}	0.40×10^{-3} mol m^{-3}

Saps collected by these techniques are generally agreed to be true samples of the sieve element contents. The velocity of exudation and the exudate concentration agree with mass transfer values obtained on intact systems (p. 56). Radioactive exudate is obtained soon after $^{14}CO_2$ is applied to the leaves and the same labelled substances are found as have been shown to move through the phloem in tracer experiments with intact plants. Exudation only takes place when the incision or aphid stylet reaches the inner conducting sieve elements.

Analytical studies of phloem exudates reveal a high dry-matter content of between 10 and 25%, of which 90% or more is sugar. In many plants sucrose is the only sugar present but other sugars, namely raffinose, stachyose and verbascose, and occasionally the sugar alcohols mannitol and sorbitol, may occur. Amino acids and amides are found in phloem saps at between 0.2 and 0.5% normally, although during leaf senescence the level may rise to 5%. A wide variety of organic and inorganic compounds, including growth substances and enzymes, are also present (see Tables 5.1 and 5.2).

5.3.1 *Sugars*

The translocated sugars are all similar, consisting of glucose and fructose units joined by a glycosidic bond as in sucrose with one or more D-galactose units attached (Fig. 5.3). The resultant sugars, raffinose, stachyose, verbascose and ajugose are sometimes found in small amounts with sucrose

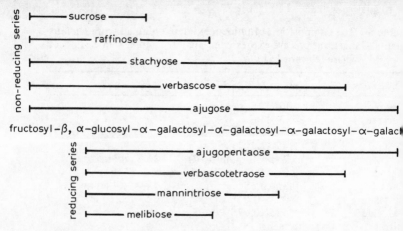

Fig. 5.3 The carbohydrates transported in the phloem are usually non-reducing members of the raffinose family of oligosaccharides. Each higher member is formed by the addition of one 1, 6-galactosyl residue.

as the major component, but in some plant families the larger oligosacchar
are more abundant (in cucurbitaceae, stachyose is the major translocate). sugar alcohols D-mannitol and sorbitol are found in some tree species, in cherry about 50% of the total dry weight of exudate consisting of sorbitol the rest being mainly sucrose. The giant brown alga *Macrocystis* translocat D-mannitol only and is the single known example of the total absence of sugars from the translocation stream [90]. In general seaweeds are known to produce large amounts of mannitol.

The hexoses, glucose, fructose and galactose are rarely found in phloem exudates although glucose and fructose (not galactose) are commonly present in phloem tissues. These monosaccharides are not translocated and occur only in non-conducting cells as products of sucrose hydrolysis.

The physiological significance, if any, of the fact that translocated sugar are always non-reducing has not been elucidated. Sucrose may be regarded as a protected derivative of glucose. Non-reducing sugars do not enter directly into cellular metabolism and may provide a more stable material for long-distance transport. Also, weight for weight, these sugars contain more available energy because some is released on hydrolysis of the glycosidic bond. A further possible advantage is that more carbon can be transported as oligosaccharides without alteration of the osmotic pressure (OP), molar sucrose having a similar OP to molar glucose, but the sucrose solution containing twice as much carbon.

5.3.2 *Proteins and amino acids*
Protein levels in phloem exudates are relatively high, ranging from 0.3 to 9.8 mg ml^{-1} (Table 5.1). The major component is P-protein but a number of enzymes, particularly those associated with carbohydrate and nitrogen metabolism are also found. The relative stability of the transported oligosaccharides is due to a complete absence of invertase in the sieve

Table 5.2 The amino acid composition of phloem exudate obtained from *Ricinus* plants. [51]

Amino acid	Concentration mol m^{-3}	%
Glutamic acid	13.0	34.76
Aspartic acid	8.8	23.53
Threonine	5.4	14.44
Glycine	2.4	6.42
Alanine	2.0	5.35
Serine	1.6	4.28
Valine	1.6	4.28
Isoleucine	1.0	2.67
Phenylalanine	0.6	1.60
Histidine	0.4	1.07
Leucine	0.4	1.07
Lysine	0.2	0.53
Arginine	Trace	Trace
Methionine	Trace	Trace
Total amino acids	35.2	100

element sap and probably to the absence of those enzymes responsible for the splitting of raffinose and stachyose in species transporting these sugars. The utilization of sucrose in the formation of sieve element starch or of callose is made possible by the presence of UDPG-fructose glycosyltransferase. Phosphofructokinase, which plays a key role in glycolysis, is also present but may be inhibited by the high ATP content (p. 53). The enzymes present in the sieve elements may be synthesized in the associated companion cells. Although protein synthesis occurs in sieve elements during their development, phloem exudates are incapable of any such synthesis.

A complex biochemical system is present within the sieve elements [39] with those enzymes concerned with starch synthesis, with glycolysis, with transaminations, and some of the enzymes of the citric acid cycle, being present. The enzyme glucose-6-phosphate dehydrogenase, which is the starting point of the pentose phosphate pathway, is also reported. These enzymes and their biochemically interrelated reactions are shown in Fig. 5.4. In addition, other investigators have reported ATPase, ATP diphosphohydrolase, cytochrome oxidase, α-D-galactosidase, peroxidase and polyphenol oxidase in the sieve elements of some other plants. The presence of this complex array of enzymes in the assimilate stream may be fortuitous or they may play an important metabolic role.

The wide variety of amino acids and amides which are found in phloem saps (Table 5.2) differs with the age and species of the plant. A selective uptake of amino acids and amides into the phloem has been proposed, this selectivity possibly varying among plant species and not simply reflecting that of the source leaf, but having a different composition.

When $^{14}CO_2$ is fed to an exporting *Ricinus* leaf, aspartic and glutamic acids, serine and alanine are the labelled amino acids which are exported as primary products of photosynthesis.

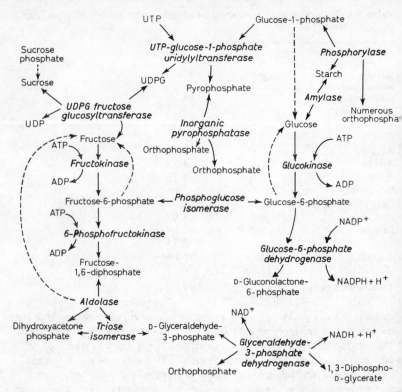

Fig. 5.4 Enzymes of the sieve tubes and their possible biochemically related reactions [135].

5.3.3 *Inorganic solutes*

A number of inorganic ions are found in phloem saps, the cationic content of which indicates the involvement of some membrane regulation; $[K^+] \gg [Na^+]$, and $[Mg^{2+}] > [Ca^{2+}]$. These cation levels probably reflect the cytoplasmic origins of phloem sap, similar cation ratios being observed in the cytoplasm of those plant cells so far analysed.

K^+ is the major cation with levels up to 112 mol m^{-3} (see Table 5.1). Such high K^+ levels increase the amount of energy which needs to be expended if K^+-driven electro-osmosis is considered as a major driving force for assimilate movement (see Section 5.2). K^+ deficiency is known to have a strong inhibitory effect upon translocation in sugar cane, observable before the net photosynthesis of the source leaves is affected [56]. The high Mg^{2+}/Ca^{2+} provides a suitable environment for a motile machinery such as that involved with streaming in *Nitella*, where Mg^{2+} promotes and Ca^{2+} inhibits the streaming process [13].

Inorganic anions are present in much lower amounts than the cations, the balance being provided by organic anions. Cl^- and PO_4^{3-} are the most common inorganic anions with traces of SO_4^{2-} and HCO_3^- often present. NO_3^- is universally absent from phloem saps, the nitrogenous materials

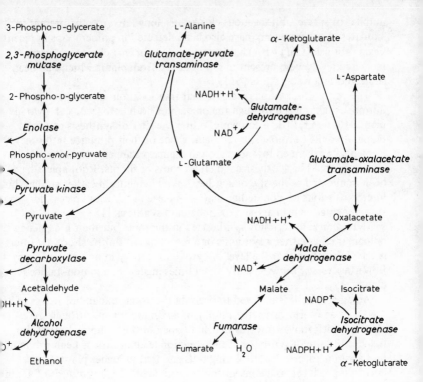

Fig. 5.4 (*continued*)

being totally organic. In those species which transport NO_3^- in the xylem, the absence of NO_3^- in the phloem sap argues against any *direct* circulation of nitrogenous material between the two conducting pathways. A pathway undoubtedly exists, with the cells of the source providing an indirect link.

In many species the levels of the phloem mobile ions mentioned above increases markedly during leaf senescence when they are exported through the sieve tubes prior to leaf fall. In addition to the major inorganic ions outlined above, traces of Mo, Cu, Fe, Mn and Zn have also been recorded in *Yucca* phloem exudate [124].

5.3.4 *Growth substances*

The existence of growth substances or hormonal factors in the phloem sap has been the subject of considerable speculation and controversy. The existence of endogenous auxin in phloem sap was first demonstrated in *Fagus* and has since been found in *Salix* and in *Ricinus*. That the observed auxin activity is endogenous IAA (indole-3-acetic acid) has been verified by mass spectroscope analysis on a sample obtained from one litre of *Ricinus* phloem sap [52].

Gibberellin-like activity has been reported in the phloem saps of a number of species. Endogenous cytokinins, long known to be present in exudates of xylem origin, have also been detected in phloem saps in recent years. Abscisic acid (ABA) has been found in *Ricinus* phloem exudate: increased levels were present in sap obtained from plants which had been subjected to a water stress [61].

The above observations indicate that translocation of endogenous hormones takes place through the phloem of both herbaceous and woody plants. However, little is known of either the sites of synthesis of these phloem-mobile hormones or the significance of their presence in the sieve elements. As these endogenous growth regulators are also found in xylem saps they could be synthesized in either roots or shoots. Root apices have been implicated as the site of synthesis of gibberellins in earlier studies, but interconversions of gibberellins may take place in the root apices, the shoot system being the primary site of gibberellin synthesis [18].

Cytokinins are certainly synthesized in the roots but their presence in the phloem implies either a synthesis or a recycling within the shoot. Of interest is the observation that the level of cytokinin activity in phloem sap from flowering *Xanthium* was nearly seven times higher than in non-induced plants [95].

Auxin synthesis is believed to occur in the young expanding leaves of the shoot apex. However, the finding of auxin movement through the phloem which moves towards the shoot apex, makes it unlikely that this tissue is the site of synthesis of phloem-mobile auxin. It has been proposed that auxin is formed by autolysing cells and that in higher plants it is normally produced as a consequence of cell death. Thus synthesis of auxin may occur as a result of the autolysis of sieve elements or of autolytic processes occurring during differentiation.

Changes in the concentration of these hormones in the phloem of photo periodic plants grown under different daylengths support the idea that a stimulus may be transmitted by differences in the balance between growth promotors and growth inhibitors in the phloem sap. There is a vital need for further studies of phloem-mobile hormones, both to elucidate the sites of their synthesis within the plant and the role of these substances in the control of assimilate distribution and subsequent growth and development.

5.3.5 *Other physiologically important substances*

In addition to the sugars and other substances described above other physiologically important materials occur in phloem saps. Organic acids have been found in those species so far analysed. Kursanov [72] proposed that most of the organic acids formed in the leaf tissues are excluded from the sieve elements, only small quantities of malic and citric acids being selectively transported in rhubarb. Citric, tartaric and possibly oxalic acids were found in aphid stylet exudate from *Salix* [93]. Malic acid, which is the most common of the acids of the TCA cycle to accumulate within plant cells, has been reported to move in considerable quantities in certain seedling stages of the soybean [88] and is the only organic acid found in *Ricinus* exudate [51]. Malic acid has also been reported in substantial amounts in the phloem exudate from the fruits of certain members of the

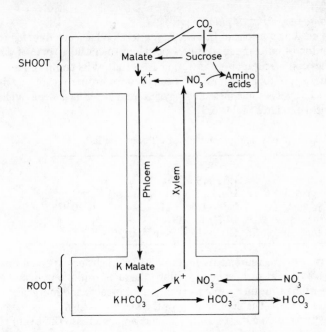

Fig. 5.5 A positive feedback regulatory mechanism for the control of NO_3^- uptake [15].

Fabaceae [92] where organic acids contributed some 2.5% of the ^{14}C activity recovered from the exudate of plants fed $^{14}CO_2$. Malate has been implicated in the regulation of NO_3^- uptake in a positive feedback regulatory mechanism [15] as illustrated in Fig. 5.5.

ATP has been identified in phloem saps at relatively high concentrations (Table 5.1). This factor is seldom included in published calculations of the energy available for translocation, the respiratory activity of the bark usually being considered as the only source of energy. Other organic phosphates have also been found in phloem saps.

The nucleic acids DNA and RNA have been reported [137] and the vitamins ascorbic acid, biotin, nicotinic acid, pantothenic acid, folic acid, pyridoxine, riboflavin and thiamine are also found [136].

Phloem saps can thus be seen to contain, at least in small amounts, a large number of the myriad of substances which are involved in plant metabolism. The list of substances in phloem exudates continues to grow as analytical techniques become more sophisticated and it seems probable that at least traces of all the water-soluble substances which are naturally occurring within plants will be found. In addition, almost any substance applied exogenously will be translocated to some extent within the phloem. Readers are referred to Crafts and Crisp [30] for a detailed discussion of the movement of exogenous substances.

5.3.6 *Some properties of phloem sap*

In the preceding sections of this chapter, attention has been directed to the composition of phloem sap. It is now pertinent to consider some of the properties conferred upon this solution as a result of its composition. Previously approximate values have been used for properties such as viscosity, but in recent years some measured physical properties have been published for phloem exudates (Table 5.3).

Table 5.3 Some physical properties of phloem exudate

pH	7.2−8.5
Conductance	1.32 mS m^{-1} at 18°C
Osmotic pressure	14.2−18.3 bar ($\times 10^5$ Pa)
Viscosity	1.34−1.58 $\times 10^{-3}$ N s m^{-2} at 20°C
Density	1.059 at 20°C
Surface tension	5.03 $\times 10^{-2}$ N m^{-1} at 20°C

The viscosity measurements are those which might be expected for a sucrose solution at such a concentration. Measured osmotic pressures are also consistent with the large concentration of organic and inorganic solutes present in the sap. The conductivity of the exudate is high, as would be predicted with such a large quantity of charged materials. The pH of phloem exudate is always slightly alkaline, a reflection of the high levels of K^+ present (p. 50). The ionic balance in the phloem sap is maintained by organic anions such as malate and the amino acids, the greater proportions of which are dicarboxylic.

Changes in both temperature and concentration have marked effects on viscosity and it is therefore likely that the viscosity of the sieve tube sap does not remain constant but fluctuates according to the changes in the physiological conditions.

5.4 Movement in the phloem

Translocation of assimilates is normally from the sites of production in the photosynthetic tissues (sources) to the sites of utilization or storage in other parts of the plant (sinks), at a rate proportional to the gradient of concentration in the phloem.

5.4.1 *Translocation rates and mass transfer*

The rapid growth of fruits and storage organs indicates that considerable amounts of material are transported to them. Early workers attempted a quantitative assessment of this transport by measuring the dry weight increase of fruits or tubers, estimating the cross-sectional area of the phloem tissue through which the transport takes place and then calculating the velocity with which a solution of a certain concentration would have to flow. Many difficulties are inherent in this method: corrections must be made for any movement of materials in the reverse direction, for losses due to respiration and, if the organ is photosynthetic, for the local synthesis of metabolites. These other components have not or can not always be

Table 5.4 Measured velocities for the translocation of assimilates through the phloem [30]

Plant	Velocity (m h^{-1})
Glycine max	0.17 − (72.00)$^+$
Salix sp.	0.25 − 1.00
Cucurbita pepo	0.30 − 0.88
Gossypium hirsutum	0.35 − 0.40
Heracleum mantegazzianum	0.35 − 0.70
Metasequoia glyptostroboides	0.48 − 0.60
Beta vulgaris	0.50 − 1.35
Phaseolus vulgaris	0.60 − 0.80
Ricinus communis	0.84 − 1.50
Triticum aestivum	0.87 − 1.09
Cucurbita melopepo torticollis	2.50 − 3.00
Saccharum officinarum	3.00 − 3.60

$^+$ The high rate of 72 m h^{-1} reported for Glycine max is probably not due to transport through the phloem.

directly measured and have to be estimated indirectly, which necessitates a number of assumptions and inevitably leads to errors.

One of the earliest such estimates was performed on a potato tuber [33]. 50 g of carbohydrate accumulated in 100 days through a stolon with cross-sectional area of phloem of 0.422 mm^2, required a flow rate of about 0.4 m h^{-1}, assuming a 10% sugar solution flowing unidirectionally. This simple calculation presumes a cylinder of liquid moving at a certain speed, although in practice such flow is always paraboloid; thus the peak velocity would be twice the average velocity as calculated above. However, it is remarkable how well this simple estimate agrees with velocities calculated in more recent times using sophisticated techniques (see Table 5.4).

A number of investigators have observed that different substances may be translocated at different velocities in the phloem, a result interpreted as demonstrating the independent movement of different substances through the phloem [77] but probably merely reflects that some substances, such as sucrose, are introduced into the sieve tubes more rapidly than others.

It is important to distinguish between the velocity of transport expressed as the distance moved by an individual molecule per unit time and the total amount of material transported in a given time. For comparative purposes it is sometimes more useful to express the translocation capacity in terms of the latter, that is the amount of carbohydrate transferred per unit time through the cross-sectional area of the phloem. This value, the 'specific mass transfer' can be calculated as follows:

$$\text{Specific mass transfer (SMT)} = \frac{\text{Transfer of dry weight per unit time}}{\text{Cross sectional area of phloem}}.$$

Using the values quoted above for the potato experiment,

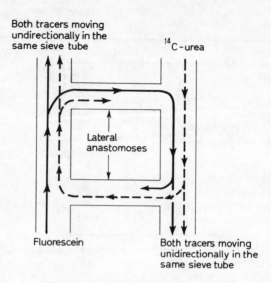

Fig. 5.6 Lateral transfer and loop paths as explanation of apparent bidirectional movement [37].

$$\text{SMT} = \frac{50}{0.422 \times 10^{-6} \times 24 \times 10^{-2}} = 4.9 \times 10^4 \text{ g m}^{-2} \text{ h}^{-1}$$

a value similar to that of a number of fruits and storage organs, the SMT of which range from about 2 to 5×10^4, with an average value of 3.6×10^4 [25]. However, these figures are based on the cross-sectional area of the whole phloem and should be increased fivefold as the sieve elements may occupy only 20% of the cross sectional area of the phloem, although this figure is based on only a few measurements [25].

5.4.2 *Direction of movement*

Sugars transported out of the photosynthetic 'source', the leaves, move in the stem both upwards and downwards towards 'sinks' such as growing points and storage organs. In some cases upward movement may involve lateral transport to the xylem and subsequent movement in that tissue, although in many experiments this possibility has been precluded and upward transport through the phloem unequivocally demonstrated. Thus in some cases substances may be translocated in opposite directions within the plant simultaneously. Whether this bidirectional movement can occur simultaneously in the same part of an individual sieve element has been the subject of numerous investigations and the answer is of fundamental importance when evaluating various postulated mechanisms of translocation.

Bidirectional movement has been demonstrated within the same phloem bundles using ^{14}C and ^{32}P as tracers [16], but whether simultaneous movement was taking place within an individual sieve tube was not resolved. Fluorescein (a phloem-mobile dye) and ^{14}C both appeared in the honeydew

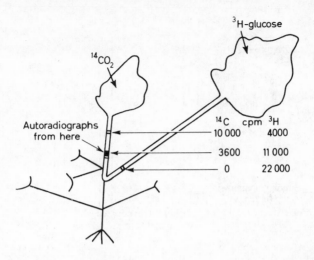

Fig. 5.7 Didirectional movement of ^{14}C and ^{3}H in a squash plant [120].

from aphids feeding on experimental plants, the fluorescein moving upwards and the ^{14}C moving downwards [37]. As aphids are known to puncture and feed on only one sieve element either bidirectional movement was taking place in a single sieve element or lateral transfer through sieve element anastomoses had created a 'loop-path' mixing the tracers moving in opposite directions in adjacent sieve elements (Fig. 5.6).

Using high resolution autoradiography bidirectional translocation within a single sieve element has been demonstrated [120]. $^{14}CO_2$ was fed to one leaf and tritiated glucose to another (Fig. 5.7) timing the feeding so that the advancing fronts of each label were localized in the section of petiole below the ^{14}C-fed leaf. Autoradiographs from this region indicated that the activity of both tracers was confined to the sieve elements and appeared in the same sieve elements, providing a strong case for bidirectional movement within a single sieve element. However, at a certain stage [42] leaves show a transition state in which they reverse from importing to exporting, suggesting that the above bidirectional movement may have occurred in this transition stage [38] (the ^{14}C-labelled leaf used in the experiments was not fully mature). Thus although bidirectional flow can occur, when the source and sink are well defined and high rates of mass transfer are observed, the translocation is essentially a one-way system.

Although there is some lateral exchange of materials between the phloem and surrounding tissues, substances moved in the phloem show a predominantly linear pattern of distribution, moving both up and down the stem in line with a supplying leaf, a pattern altered in some species by defoliation. However, considerable movement in the radial direction from phloem to xylem is known to occur through the vascular rays.

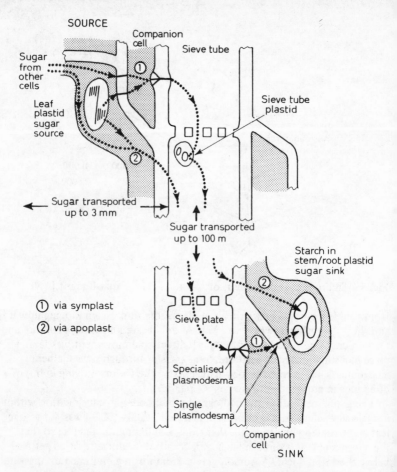

Fig. 5.8 The path of sugar translocation from manufacture in the chloroplas to storage in a stem or root plastid [131].

5.4.3 *Loading and un-loading of the transport system*

Continued movement of sugars through sieve elements is dependent on loading at the source and unloading at the sink. There are two possible pathways between photosynthetic mesophyll and sieve elements through which sugars may move. These two pathways, which exist in parallel, are the cytoplasmic continuity, the symplast, and the cell wall continuity, the apoplast. Movement through the symplast could take place through the numerous plasmodesmata connecting between the cells, the driving force for this movement being diffusion aided by cyclosis of the cytoplasm within the cells. Alternatively, or as well as, sugar movement may take place by diffusing through the apoplast of free space. Such free space transport would be driven by the active release of sugars from the mesophyll cells followed by an active accumulation into the sieve elements, a process which would provide the necessary gradient for apoplastic movement. Thus both

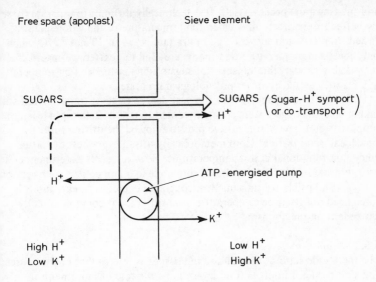

Fig. 5.9 A scheme for a proton efflux/potassium influx pump at the plasma membrane of the sieve element with proton co-transport of sugars down the resultant proton gradient [81].

parallel pathways would appear to require diffusional driving forces and as mesophyll cells are always closely associated with sieve elements only relatively small concentration differences would be required (Fig. 5.8).

An important distinction between these two possible pathways is that loading of the symplasmic pathway would be more light sensitive and respond rapidly to changes in the concentration gradient between the chloroplasts and the sieve element. Light dependence of rapid translocation has been reported for sugar cane. Active transport of sugars into and out of the apoplast would be less light sensitive and could be powered by ATP derived from respiration [5]. Unloading would be achieved in a similar manner, the concentration gradients being reversed and created by sugar utilization at the sink.

Results obtained with *Ricinus* indicate that a proton driven co-transport of sugars may take place [81]. A coupled H^+ efflux/K^+ influx pump is envisaged which establishes a proton gradient down which protons move from the free space into the sieve element, co-transporting sugar inwards (Fig. 5.9). Such a scheme is consistent with the composition of phloem sap, its high pH, high $[K^+]$, and with the action of inhibitors on loading and long-distance transport, which presumably inhibit the H^+/K^+ pump.

5.5 Physiology of the phloem
Factors affecting translocation.

5.5.1 *Temperature*
Temperature affects both the rate of synthesis and utilization of assimilates

and the transport process itself. It is technically difficult to separate these two effects experimentally. When the temperature of the whole plant is varied, translocation shows a maximum rate between 20 and 30°C; when only localized temperature treatments are used the pattern of results is remarkably similar, the transport of sugars being influenced by temperature in a manner similar to other physiological processes.

There have been reports over the past decade or so of translocation still taking place through a stem segment cooled to 0°C or below [44], lending support to the view that a passive pressure flow is the driving force for translocation as distinct from metabolically-driven processes, the latter being non-functional at low temperatures. However, such experiments must be interpreted with caution, the driving force in a non-cooled stem section may be responsible for movement through the cooled segment, and this force need not itself be temperature insensitive, i.e. it could be directly dependent on metabolism.

5.5.2 *Metabolism*
The metabolic state of the conducting tissues as well as that of the source and sink tissues is important in determining the rate of movement of assimilates through the plant. Killing a section of petiole with a jet of steam will prevent the movement of assimilates out of that leaf. Less drastic treatments such as anaerobiosis or metabolic inhibitors will also often cause a marked reduction of transport, although it is difficult to assess whether the inhibitor exerts its effect directly on the translocation process or indirectly by acting on the metabolism of the source and sink. Transloca tion may be controlled to a greater extent by the metabolism of source and sink tissues than by that of the conducting tissue itself. On the other hand number of investigators believe that metabolism is directly involved in the conducting process (see Chapter 6), and inhibition of long-distance transport by metabolic inhibitors has been observed under conditions where both source and sink are not subjected to the inhibitor [100] althou it may be that the inhibitor merely induces physical blockage of the sieve plate pores [46].

5.5.3 *Light*
There are a number of conflicting reports about the effect of light upon translocation. For example, no effect of light on translocation from leaves other than its effect on the supply of assimilates has been observed [125], whereas different distributions of the translocate between shoot and root in light and darkness are reported (Table 5.5), transport to the root being favoured over transport to the shoot in the dark. A more rapid movement of assimilates towards the root during the night has also been observed by other investigators and presumably reflects changes in the relative sizes of the shoot and root sinks under the different conditions.

In sugarcane, light influences both the polarity and the amount of trans location, both total and basipetal translocation being greater in the light than in the dark [54]. In these experiments light affected translocation at intensities which were too low for CO_2 assimilation and thus the effect of

Table 5.5 Translocation of ^{14}C-assimilates under the influence of both light and dark conditions [87]

		% Translocated	
		Stem tip	*Root*
	3 hours light	2	4.4
15 minutes $^{14}CO_2$ ↗ ↘ in light			
	3 hours dark	0.5	16.5

light was not upon the photosynthetic processes. Subsequent experiments [57] indicated that the red end of the spectrum was responsible for this light effect and that translocation may be triggered by different mechanisms in the light and in the dark.

5.5.4 *Mineral deficiencies*

Little is known about the effect of mineral deficiencies on translocation. Many deficiencies will obviously have an adverse effect on this process due to their influence on the general metabolism and growth of the plant. Some evidence of a direct role has been claimed for boron, which has been demonstrated to facilitate both the absorption and translocation of ^{14}C-sucrose. An ionizable complex is postulated between sucrose and boron which is transported across cell membranes with a greater facility than sucrose alone. However, it may be that boron enhances the activity of apical meristems which promotes translocation to these regions or that exclusion of boron from sieve elements might prevent callose formation [36]. In sugarcane, a lowering of the K^+ level which did not result in visible deficiency symptoms has been reported to decrease translocation, associated with an upset in phosphorylation [55].

5.5.5 *Concentration gradients*

Translocation always occurs from regions of high to regions of low sugar concentration, down a gradient which can be removed or reversed by defoliation, a process which also stops transport. Gradients of about 0.01 mol m^{-1} have been found in the stem of white ash, positive in the downward direction [138]. These gradients were altered on defoliation, disappearing for some sugars and becoming negative for other (Fig. 5.10). The significance of such gradients with respect to the translocation mechanism are discussed later.

5.5.6 *Hormone-directed transport*

During the last thirty years or so the concept of hormone-directed transport has developed, patterns of nutrient movement being controlled by the physiological effects of endogenous growth hormones as well as by concentration gradients (see Section 5.4). The critical problem still to be elucidated is the determination of the exact nature and degree of this hormonal control, whether a direct effect on the translocation mechanism occurs or if the effect is indirect due to enhanced growth in certain sink regions. Evidence for the latter effect has been observed by a large number

61

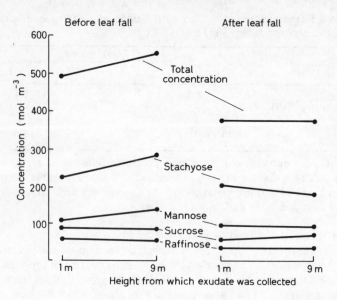

Fig. 5.10 Concentration gradients along a tree trunk of white ash before and after defoliation [138].

Table 5.6 The effect of hormones on ^{14}C-sucrose transport in beans [85]

Treatment	Mean count rate of decapitated internode	Mean fresh weight (mg) decapitated internode
Lanolin control	122	5.0
IAA	616	5.1
IAA + GA + benzyl-adenine	2016	6.8
LSD ($P = 0.05$)	322	0.5

of independent investigators who observed the accumulation of assimilates in regions of high auxin. Short-term experiments have shown the accumulation of significant amounts of ^{14}C-labelled assimilates in auxin treated stumps of various plant species. Evidence of synergism between different growth substances has been demonstrated (Table 5.6), the fast accumulatory response to IAA being enhanced by GA and kinins.

Endogenous hormones have been detected in phloem exudates (see Section 5.3) and transport of exogenously applied growth substances has been observed. The presence and long-distance transport of hormones in the phloem, when considered in the light of the above observations on the accumulatory response of treated tissues, suggests that a direct hormonal stimulation of the translocation mechanism may be possible.

Bibliography

Canny, M.J. (1973), *Phloem Translocation,* Cambridge University Press, London.
Contains useful information on specific mass transfer and of the authors' views on the translocation mechanism.

Crafts, A.S. and Crisp, C.E. (1971), *Phloem Transport in Plants,* Freeman, San Francisco.
A detailed account of the phloem and of translocation, with a strong bias towards pressure flow as the transport mechanism.

Peel, A.J. (1974), *Transport of Nutrients in Plants,* Butterworths, London.
Concentrates on phloem transport. Gives a detailed historical development of studies on mechanism, with a useful survey of aphid utilization.

Wooding, F.P.B. (1971), *Phloem,* Oxford Biology Readers, Oxford University Press.
A short concise account of phloem structure and function.

6 Driving forces for long-distance transport

6.1 Transpiration and the cohesion theory

It is now well established that the transpirational flow of water and solutes through the xylem conduits is a bulk or mass flow of solution. Ringing experiments, sap analysis, and tracer movement in intact plants have all confirmed that the upward movement of sap from the roots is through the xylem. According to the cohesion theory the driving force for this long-distance transport during transpiration is the gradient of negative hydrostatic pressure (tension) between one end of the system and the other. The negative pressures develop in the xylem of the leaves as water loss by transpiration reduces the leaf water potential. As a result of liquid phase continuity, and the fact that the osmotic pressure of the xylem sap is very small, the water potential of the xylem will be nearly equal in magnitude to the water potential of the leaf, resulting in a negative pressure in the xylem. This negative pressure is transmitted through the liquid continuity within the xylem between leaves and root, the breaking of the liquid columns being prevented by cohesion between adjacent water molecules and adhesion between the water molecules and the xylem wall.

This liquid flow through the xylem may be quantified. For narrow cylindrical tubes such as xylem vessels the Hagen-Poiseuille relationship is appropriate:

$$\phi_v = \frac{r^2}{8\eta} \Delta P \tag{6.1}$$

where ϕ_v is the volume flow, r the radius, η the viscosity, and ΔP the hydrostatic pressure gradient.

63

How well does the above equation describe sap flow through the xylem ϕ_v in the xylem of a transpiring tree may equal 1.0 mm s^{-1} [110]. For a xylem element 20μm radius the required pressure gradient would be 0.2 bar m^{-1}. An additional 0.1 bar m^{-1} of pressure gradient due to gravity exists even in the absence of any flow. Thus a gradient of 0.3 bar m^{-1} wou suffice for conduits of 20 μm radius. For larger conduits this same pressure gradient would result in a greater ϕ_v. Measured pressure gradients range between 0.1–0.5 bar m^{-1} [138] indicating that flow through xylem conduits closely approximates the Hagen-Poiseuille relationship. However, a survey of the water conducting efficiency of various plants (Table 6.1)

Table 6.1 The efficiency of water conductivity in various plants [64]. Comparison of observed (v_0) and predicted (v_t) rates of water movement predicted from the Hagen-Poiseuille relationship

Plant	Conductive efficiency $v_0/v_t \times 100$
Vitis vinifera	100
Aristolochia sipho	100
Atragene alpina	100
Root wood of oak	84
Root wood of beech	37.5
Helianthus annuus	32
Rhododendron ferrugineum	20
Rhododendron hirsutum	13

indicates that whereas the wood of climbing lianas agree precisely and ring porous species such as oak give close to theoretical values, diffuse porous species such as beech are considerably below those predicted. Such deviations are not too surprising, however, when the presence of end walls in xylem tracheids and vessels is taken into account. Short vessels and tracheids will obviously show conductivities less than those calculated for open-ended cylinders.

The magnitude of the tension in the xylem will be equivalent to the water potential values in the leaf tissues and therefore tensions of −100 ba or lower can presumably exist in some plants. Theoretically the maximum tensile strength of water is equivalent to about −18 000 bar, but in many plants it is probable that cavitation occurs in most of the vessels before values of −100 bar are reached, and will occur in the larger vessels at less negative values. The production of sound by cavitating water columns has been detected in both physical systems and in xylem vessels. In some herbaceous plants sound-detected cavitation has been observed to occur at relatively small tensions of −5 to −15 bar [84], suggesting that cavitatio is a frequent occurrence. As the plants usually recover during the hours of darkness it is possible that root pressure (see p. 26) may have a physiologic role in restoring the cavitated water columns, particularly in herbaceous plants where root pressures are considerable in proportion to their height.

It is difficult to simulate the tensions which occur in the xylem in physical systems, without cavitation taking place. Water in the xylem is well-filtered by the root, thus preventing the entry of small particles which can serve as nuclei for bubble formation. Also the narrow tubes of the xylem with wettable walls are ideal for supporting negative pressures. When water in the xylem is under tension the turgor of living bark cells in equilibrium with the xylem undergoes changes, usually showing a diurnal pattern, which can be detected as variations in the trunk diameter of trees using a dendrograph [139].

The velocity of the transpiration stream has been measured using a heat pulse method which employs localized heating of the xylem sap together with thermocouples along the stem. The transpirational responses are more rapid in twigs and small branches (Fig. 6.1) an observation consistent with the cohesion theory. Using the heat pulse method it has been observed that

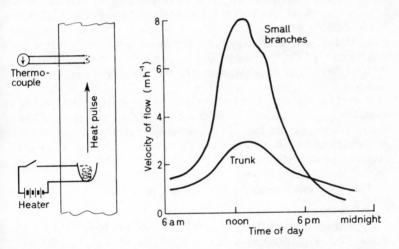

Fig. 6.1 Heat pulse method of determining xylem sap flow rate [139].

water movement through the xylem is not interrupted at temperatures in the vicinity of $0°C$, provided that no actual freezing occurs [140]. This observation indicates that sap movement through the xylem is not directly dependent upon metabolic processes, and is therefore in accord with the cohesion theory.

Although now widely accepted the cohesion theory has in the past been the subject of considerable controversy. In most respects it explains how water can rise through the xylem of tall trees above the height equivalent to zero atmospheric pressure (approximately 10 m), but for its operation it requires continuous water columns. However, flow continues even when a large proportion of the larger xylem vessels are air-filled [99] and deep overlapping lateral cuts into the stem do not stop xylem sap movement, which continues at a reduced rate [98]. These observations are not necessarily inconsistent with the requirement of the cohesion theory. Only a small proportion of the total xylem vessels are necessary at any one time for the transpirational requirement and overlapping cuts can be bypassed laterally, maintaining liquid phase continuity. Cavitation of water columns will undoubtedly occur, but the resulting air embolism is isolated within the cavitated elements and is unable to spread laterally. This is because the pit pairs between one cell and the next are effectively closed by the pressure difference. Although pores exist through the pit membrane these do not exceed 0.01 μm radius and thus pressure gradients of at least 150 bar would be required to pull the air-water interface through the pit membrane. As indicated above, such high tensions are unusual.

Another problem related to cavitation, and still not adequately explained is how the xylem vessels of tall trees refill once they have been ruptured. Root pressure is undoubtedly sufficient for herbaceous plants and small trees, but the osmotic pressure of xylem sap rarely exceeds 2 bar which would support a maximum sap rise of 20 m, assuming the root system behaved as a perfect osmometer. It is possible that the vessels of tall trees do not refill once ruptured and that continued growth of new xylem elements is required. In general it is only the xylem elements formed by recent growth which are fully filled with water. There is also some evidence that water columns may be rapidly re-established by the re-dissolving of gas bubbles within ruptured xylem vessels [84].

It would appear that the mechanism of the long-distance transport of solutes through the xylem of plants has been satisfactorily resolved. This is in direct contrast to the study of solute transport through the phloem where as discussed in the following sections, attempts to resolve the mechanism have been unsuccessful.

6.2 Postulated mechanisms for phloem transport

6.2.1 *Pressure-driven mass flow*

The concept of pressure flow is generally attributed to Münch [86] although it was in fact originally postulated in the earlier work of Hartig [53]. In its simplest form, a unidirectional mass flow of solution is envisaged to take place through the sieve elements, driven by a pressure gradient established by a high turgor pressure at the source and a lower turgor pressure at the sink. Photosynthetic products will cause a raised osmotic pressure at the source, in the leaves, and the subsequent osmotic entry of water will create a high turgor pressure. The removal of sugar from the phloem in the sink area and the resultant osmotic loss of water from the sieve elements will lower the turgor pressure. It is envisaged that the water which moves out at the sink is transported to the leaves once again via the xylem although such a circulation of water need not occur if sufficient water is removed from the system at one end of the plant or the other as a result of growth (Fig. 6.2).

The pressure-flow theory is very attractive in its basic simplicity and is supported by a considerable volume of physiological evidence (see Section 5.4). However, the structure of the sieve element is not apparently suited to a pressure-driven mechanism as the major motive force for assimilate movement through the phloem. The necessary driving force may be calculated using the Hagen-Poiseuille relationship (p. 63). When measured and guestimated values are substituted, for example 1.0 m h^{-1} for

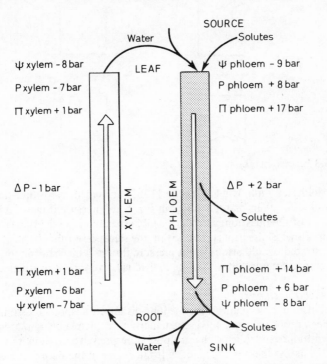

Fig. 6.2 Hypothetical model of pressure flow through the phloem. Water potential (ψ) equals hydrostatic pressure (P) minus osmotic pressure (π), water always moving in response to $\Delta \psi$. In both xylem and phloem it is ΔP which is postulated as the major driving force.

ϕ_v, 1.5×10^{-3} N s m^{-2} for η (the value for a 10% sucrose solution) and 12 μm for r (willow sieve tubes) a gradient of 0.23 bar m^{-1} is obtained [126]. This ignores the sieve plates which have been calculated to require an additional 0.32 bar m^{-1} using the same Hagen-Poiseuille relationship. Thus the total required gradient is about 0.6 bar m^{-1} for a sieve tube with completely open pores. A modified calculation has been proposed [30] which reduces the value of the required pressure gradient slightly, but the conclusion from such calculations is that pressure flow could provide the sole motive force for flow only through sieve elements with open sieve plate pores. In very tall species such as *Sequoia* and the giant eucalypts, which attain heights of 100 m, it is necessary to postulate reduced velocities of flow, larger sieve pores or a more concentrated phloem sap to accommodate the acceptability of a pressure flow mechanism operating in these tall trees.

As more evidence accumulates that the sieve plate pores are not open, but are at least partially occluded by P-protein, pressure flow seems increasingly improbable. P-protein filaments 10 nm in diameter spaced 20 nm apart would increase the pressure required to 0.14 bar across each sieve plate for flow at 1.0 m h^{-1} through sieve plates 1 μm thick. Given a minimal value of 2000 sieve plates m^{-1}, the ΔP would have to be 280 bar m^{-1},

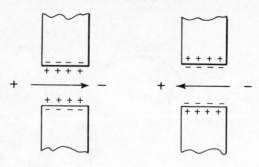

Fig. 6.3 Electro-osmosis.

a physiologically impossible pressure [126]. Thus if the sieve tube has any structure other than that illustrated in Fig. 5.2b, pressure flow would appear to be out of the question, all other possible structures being inimical to pressure flow. Bidirectional flow within the same sieve tube, if unequivocally demonstrated, would also rule out pressure flow as a mechanism, unless some compartmentation within the individual sieve tube is postulated.

6.2.2 *Electro-osmosis*

Electro-osmotic mechanisms for the long-distance transport of assimilates have been proposed, [40, 113], the underlying principle being that water is induced to move through a charged pore in response to a gradient of electric potential across the pores (Fig. 6.3). If the charge in the pore is the result of fixed negative charges then water will move towards the negative pole of the gradient. Conversely, fixed positive charges within the pore will result in a flow of water towards the positive pole.

Applying this principle to the sieve element it has been proposed that the sieve plate pores provide channels for electro-osmosis, the resultant flow of water dragging the sucrose molecules along with them. This coupling of flows enables an uncharged molecule, the sucrose, to be moved in response to a gradient of electrical potential. Fensom [40] proposed that the electrical gradient was established by the production of carbonic acid by respiration, the H^+ diffusing through the pore faster than the HCO_3^-, thus creating a gradient of electrical potential; whereas Spanner [113] ascribed the potential difference to the rapid circulation of K^+, being removed on the electronegative side of the sieve plate, as illustrated in Fig. 6.4.

Both investigators have subsequently revised their original theories, Fensom finally rejecting electro-osmosis as a driving mechanism after a detailed thermodynamic analysis of the necessary transport coefficients [123]. Spanner has not accepted the thermodynamic arguments and has proposed that the P-protein blocking the sieve pores is arranged in a structured manner (see Fig. 5.2f) so that electro-osmosis through the sieve pore is facilitated, narrower channels giving greater efficiency. There is only limited evidence available in support of an electro-osmotic movement in the phloem. An electrical potential has been demonstrated across the sieve plates [19] although it is not known whether the material was concomitantly

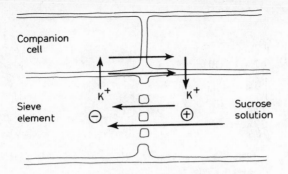

Fig. 6.4 Electro-osmotic flow of water and assimilates through the pores of phloem sieve plates [113].

translocating and, if so, the direction of flow. The measured potential may merely reflect a cutting artefact, displacement of the sieve element contents blocking the sieve plate pores which, if the proteinaceous contents carry a net charge, would result in an electrical potential gradient across the pore.

The concept of electro-osmosis as a driving mechanism for solute transport in the phloem has been criticized and rejected on the basis of a number of theoretical objections [79]. Both anions and cations are transported through the phloem (see Table 5.1), while Spanner's model would permit only cations to be transported. Considerable energy would need to be expended to maintain the circulation of K^+, in some cases more than the total energy available to the phloem as a whole, although with regard to this criticism it is worth noting that the large amount of ATP present in phloem sap was not considered in the calculations of available energy. The calculated flux of K^+ required to drive sucrose transport is some four orders of magnitude greater than the fluxes normally observed in other plant material. It has been suggested that the plasmalemma lining the side wall of the sieve tube has a 'brush border' structure which may increase the membrane area by about two orders of magnitude thus reducing the trans-membrane flux of K^+ to about 100 times higher than that of the measured values in other plant cells [114]. However, most other investigators of phloem structure have rejected the 'brush border' interpretation, the observed structure probably representing obliquely sectioned stacks of endoplasmic reticulum cisternae (Section 5.2.3). In addition to the above considerations, if bidirectional flow in the same sieve element is shown to occur, the electro-osmotic mechanism is untenable in its present form.

An alternative mechanism invoking electro-osmosis has been proposed [60] which envisages electro-conformational waves being propagated along P-protein filaments. These waves would create an electro-osmotic force when they passed through the sieve pore and this would facilitate pressure flow if acting in the same direction as the hydrostatic gradient. In-between these waves some electro-osmotic movement may occur in the opposite direction to the pressure flow, resulting in a small component of bidirectional movement. This theory supposes that the P-protein filaments are capable of

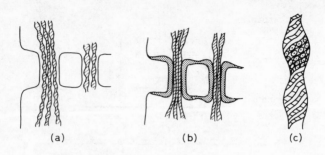

Fig. 6.5 Diagram of possible arrangement of axial strands for a micro-peristaltic model; (a) without callose, (b) with callose restricting flow. (c) Possible structure of axial strand [41].

electro-conformational changes, the possibilities of which are considered in the following section.

6.2.3 *Motile systems in sieve tubes*

The demonstration of the presence of P-protein in the sieve elements of angiosperms has led to considerable speculation as to its function in translocation. A number of investigators have proposed that the filaments may represent a motile machinery which generates a force, drawing an analogy between actin-like contactile protein and the P-protein filaments. There are several proposals for the organization of P-protein into a contractile machine for driving translocation and these will be briefly considered here.

Earlier ideas of transport through transcellular strands have been developed to incorporate a peristaltic pumping system, it being envisaged that the outer wall of the strand is composed of contractile protein filament which undergo a rhythmic contraction and relaxation resulting in a peristaltic-type wave which drives the contents through the tubules. A biophysical analysis [1] has indicated that such a system is possible on the basis of the energy requirement and flow velocities, but fine structural evidence supporting the existence of the proposed tubules is not available and many investigators doubt if such structures are present within the sieve tubes.

A micro-peristaltic movement of contractile P-protein has been proposed to take place in axial strands extending through the sieve elements (Fig. 6.5). This postulated movement activates the solution within the hollow centres of the microfilaments to generate a pulse-type flow which at the same time causes a mass flow of the solution around the microfilaments. A small surface-layer transport is also envisaged, resulting in a very rapid (10 m h^{-1}) movement along the surface of both living and non-living tissues in the phloem [41]. However, once again fine structural evidence for the existence of such filamentous strands is not available and the possibility of such a system remains not proven.

It has been suggested that P-protein is similar to the filaments which compose the fibrils responsible for generating the motive force for cytoplasmic streaming in *Nitella* [79], the P-protein envisaged as being arranged longitudinally; the correct polarity being determined by a concomitant pressure flow. Thus the force-generating P-protein is here envisaged as a supplementary mechanism which enhances the pressure-driven mass flow of solutes through the sieve tubes.

These theories are all based on the premise that P-protein has contractile properties and is chemically related to actin filaments, the basic unit of contractile protein systems. Attempts to characterize any contractile properties in the P-protein have proved mainly fruitless [129] although some isolated reports do partly support this proposed role within the phloem [119].

6.2.4 *Some concluding remarks*

The preceding discussion on possible mechanisms of phloem transport has indicated the great divergencies of opinion which exist among investigators in this field. Indeed, it has been said in a review of this subject that it sometimes appears as though there are as many concepts as there are workers [141]. However, any proposed mechanism of transport through the sieve elements must satisfy the following criteria:

(1) It must be consistent with physical laws of surface tension, flow, etc.
(2) It must explain the known physiological characteristics of the system.
(3) It must be compatible with sieve element structure.

As yet, no one hypothesis satisfies all three criteria. The problem hinges on the known structure of the sieve element which precludes rather than complements the known physiology of the system, suggesting that there is a basic error or gap in our knowledge [126]. This situation is reminiscent of Gray's paradox about the swimming of whales. Using physical models of the structure of whales it was calculated from the hydrodynamics of the system that, with the energy available from respiration, the maximum rate at which the whale could swim would be 15 knots. However, speeds of up to 20 knots are a well-established observation. Thus we have the paradox of the whale swimming faster than is physically possible according to the model. Obviously the model was wrong and it was subsequently shown that because whales alter their shape while swimming this alters the hydrodynamics of the system. Analogy with flow through sieve elements suggest that if the observed structures are interpreted in a dynamic way the resistance to flow could alter if the P-protein was redistributed when flow was taking place through the system. It appears that the P-protein is fixed in some manner to the parietal layer of the sieve element (p. 44). If, when flow is taking place, all the P-protein is displaced parietally then a low resistance pathway would be available for pressure flow. Cessation or slowing down of flow would result in the P-protein becoming redistributed in a more even manner through the lumen of the sieve element. In the event of a sudden pressure release the P-protein would be torn from its parietal fixing points by the rapid flow and block the sieve plate pores (p. 44). This hypothesis therefore supports a blockage function for the P-protein.

It may be argued that the pressure flow mechanism is too simplistic a concept by those who propose electro-osmotic, peristaltic, or other P-protein generated mechanisms of flow. However, in many ways the pressure flow mechanism is the one most consistent with the known physiology of the phloem and has been rejected primarily on structural grounds. If it can be demonstrated that the structure of a living, translocating sieve element has a central lumen free of P-protein filaments then the paradox of ploem transport may be resolved.

In the words of an old Zen saying:

'When one is very young and knows a little, mountains are mountains, water is water, and trees are trees.

When one has studied and has become sophisticated, mountains are no longer mountains, water is no longer water, and trees are no longer trees.

When one thoroughly understands, mountains are again mountains, water is again water, and trees are again trees.'

Bibliography

Aranoff, S., Dainty, J., Gorham, P.R., Srivastava, L.M. and Swanson, C.A. (eds) (1975), *Phloem Transport,* Plenum Press, New York and London.
Proceedings of a conference with many valuable papers on current controversies.

Slatyer, R.O. (1967), *Plant-Water Relationships,* Academic Press, New York and London.
The classic book on plant-water relationships. Chapter 7 is about water movement through the plant.

Zimmermann, M.H. and Milburn, J.A. (eds) (1975), *Encyclopedia of Plant Physiology,* N.S. Volume 1. Springer-Verlag, Berlin, Heidelberg, New York.
An excellent multi-author treatise on all aspects of ploem transport.

References

[1] Aikman, D.P. and Anderson, W.P. (1971), *Ann. Bot.* **35**, 61–72.
[2] Allaway, W.G. (1973), *Planta*, **110**, 63–70.
[3] Anderson, R. and Cronshaw, J. (1969), *J. Ultrastruct. Res.*, **29**, 50–59.
[4] Anderson, W.P. (ed) (1973), *Ion Transport in Plants*, Academic Press, London and New York.
[5] Anderson, W.P. (1974), *Symp. Soc. Exptl. Biol.*, **28**, 63–85.
[6] Anderson, W.P. (1975), in [11] p. 231.
[7] Atkinson, M.R. *et al.* (1967), *Aust. J. Biol. Sci.*, **20**, 589–599.
[8] Baker, D.A. (1968), *Planta*, **83**, 390–392.
[9] Baker, D.A. (1973), *Planta*, **112**, 293–299.
[10] Baker, D.A. and Hall, J.L. (1973), New Phy, **72**, 1281–1291.
[11] Baker, D.A. and Hall, J.L. (eds) (1975), *Ion Transport in Plant Cells and Tissues*, North Holland, Netherlands.
[12] Baker, D.A. and Weatherley, P.E. (1969), *J. Exptl. Botany*, **20**, 485–496.
[13] Barry, W.H. (1968), *J. Cell Physiol.* **72**, 153–160.
[14] Bell, C.W. and Biddulph, O. (1963), *Plant Physiol.* **38**, 610–614.
[15] Ben Zioni, A., *et al.* (1971), *Physiologia Plantarum*, **24**, 288–290.
[16] Biddulph, O., and Gory, R. (1960), *Plant Physiol.* **35**, 689–695.
[17] Bollard, E.G. (1960), *Ann. Rev. Plant Physiol.* **11**, 141–166.
[18] Bowen, M.R. and Wareing, P.F. (1969), *Planta*, **89**, 108–125.
[19] Bowling, D.J.F. (1969), *Biochim. Biophys. Acta*. **183**, 230–232.
[20] Bowling, D.J.F. (1973), *J. Exptl. Botany*, **24**, 1041–1045.
[21] Bowling, D.J.F. (1973), in [4] p. 483.
[22] Bowling, D.J.F. (1976), *Uptake of Ions by Plant Roots*. Chapman and Hall, London.
[23] Bowling, D.J.F. (1976), *Nature*, Lond. **262**, 393–394.
[24] Bowling, D.J.F. *et al.* (1966), *J. Exptl. Botany*, **17**, 410–416.
[25] Canny, M.J. (1973), *Phloem Translocation*, Cambridge University Press, London.
[26] Clarkson, D.T. (1974), *Ion Transport and Cell Structure in Plants*, McGraw-Hill, Maidenhead.
[27] Clarkson, D.T. *et al.* (1971), *Planta*, **96**, 296–305.
[28] Collins, J.C. and Kerrigan, A.P. (1973), in [4] p. 589.
[29] Crafts, A.S. and Broyer, T.C. (1938), *Amer. J. Botany*, **25**, 529–535.
[30] Crafts, A.S. and Crisp, C.E. (1971), *Phloem Transport in Plants*, Freeman, San Francisco.
[31] Cronshaw, J. (1974), in Robards, A.W. (ed) *Dynamic Aspects of Plant Ultrastructure*. p. 391 McGraw-Hill, Maidenhead.
[32] Cronshaw, J. and Esau, K. (1967), *J. Cell. Biol.* **34**, 801–816.
[33] Dixon, H.H. and Ball, N.G. (1922), *Nature*, **109**, 236–237.
[34] Dunlop, J. and Bowling, D.J.F. (1971), *J. Exptl. Botany*, **22**, 434–444.
[35] Epstein, E. (1973), *Int. Rev. Cytol.* **34**, 123–168.
[36] Epstein, E. (1973), *Experentia*, **29**, 133–134.
[37] Eschrich, W. (1967), *Planta*, **73**, 37–49.

73

[38] Eschrich, W. (1975), in [142] p. 245.

[39] Eschrich, W. and Heyser, W. (1975), in [142] p. 101.

[40] Fensom, D.S. (1957), *Can. J. Botany*, **35**, 573−582.

[41] Fensom, D.S. (1972), *Can. J. Botany*, **50**, 479−497.

[42] Fritz, E. (1973), *Planta*, **112**, 169−179.

[43] Fujino, M. (1967), *Sci. Bull. Fac. Educat. Nagasaki Univ.* **18**, 1−47.

[44] Geiger, D.R. and Sovonick, S.A. (1975), in [142] p. 256.

[45] Gerson, D.F. and Poole, R.J. (1971), *Plant Physiol,* **48**, 509−511.

[46] Giaquinta, R. and Geiger, D.R. (1977), *Plant Physiol.* **59**, 178−180.

[47] Ginsburg, H. and Ginsburg, B.Z. (1970), *J. Exptl. Botany*, **21**, 593−6

[48] Gunning, B.E.S. and Pate, J.S. (1969), *Protoplasma*, **68**, 107−133.

[49] Hall, J.L. and Baker, D.A. (1975), in [11] p. 39.

[50] Hall, J.L. and Baker, D.A. (1977), *Membranes and Ion Transport*, Longman, London.

[51] Hall, S.M. and Baker, D.A. (1972), *Planta*, **106**, 131−140.

[52] Hall, S.M. and Medlow, G.C. (1974), *Planta*, **119**, 257−261.

[53] Hartig, T. (1860), *Allgem. Forst-u-Jagdztg,* **36**, 257−261.

[54] Hartt, C.E. (1965), *Plant Physiol.* **40**, 718−724.

[55] Hartt, C.E. (1969), *Plant Physiol.* **44**, 1461−1469.

[56] Hartt, C.E. (1970), *Plant Physiol.* **45**, 183−187.

[57] Hartt, C.E. and Kortschak, H.P. (1967), *Plant Physiol.* **42**, 89−94.

[58] Hay, R.K.M. (1974), in Kolek, J. (ed) *Structure and Function of Primary Root Tissues,* p. 329, Slovak Academy of Science.

[59] Heine, H. (1885), *Ber. Deutsch. Bot. Ges.* **3**, 189−194.

[60] Hejnowicz, Z. (1970), *Protoplasma*, **71**, 343−364.

[61] Hoad, G.V. (1973), *Planta*, **113**, 367−372.

[62] Hope, A.B. (1971), *Ion Transport and Membranes*, Butterworths, London.

[63] House, C.R. and Findlay, N. (1966), *J. exptl. Botany*, **17**, 344−354.

[64] Huber, B. (1956) in Ruhland, W. (ed) *Hanb. Pflanzenphysiol.* Vol. III p. 511, Springer-Verlag, Berlin.

[65] Humble, G.D. and Hsiao, T.C. (1969), *Plant Physiol.* **44**, 230−234.

[66] Humble, G.D. and Raschke, K. (1971), *Plant Physiol.* **48**, 447−453.

[67] Isermann, K. (1971), *Z. Pflanzenr Bodenkunde*, **128**, 195−207.

[68] Jacoby, B. (1965), *Physiol. Plant,* **18**, 730−739.

[69] Jaffe, M.J. (1970), *Plant. Physiol.* **46**, 768−777.

[70] Jeschke, W.D. and Simonis, W. (1969), *Planta*, **88**, 157−171.

[71] Koukkari, W.L. and Hillman, W.S. (1968), *Plant Physiol.* **43**, 698−70

[72] Kursanov, A.L. (1963), *Advan. Botan. Res.* **1**, 209−274.

[73] Laties, G.G. (1969), *Ann. Rev. Plant Physiol.* **20**, 89−116.

[74] Laüchli, A. (1972), *Ann. Rev. Plant. Physiol.* **23**, 197−218.

[75] Lüttge, U. (1975), in [11] p. 335.

[76] Lüttge, U. and Pitman, M.G. (eds) (1976), *Encyclopedia of Plant Physiology* N.S. Volume 2A, Springer-Verlag, Berlin, Heidelberg, New York.

[77] Lüttge, U. and Pitman, M.G. (ed) (1976), *Encyclopedia of Plant Physiology* N.S. Volume 2B, Springer-Verlag, Berlin, Heidelberg, New York.

[78] Macallum, A.B. (1905), *J. Physiol.* **32**, 95−118.

[79] MacRobbie, E.A.C. (1971), *Biol. Rev.,* **46**, 429−481.

[80] MacRobbie, E.A.C. (1971), *Ann. Rev. Plant Physiol.* **22**, 75−96.

[81] Malek, F. and Baker, D.A. (1977), *Planta*, **135**, 297−299.

[82] Mason, T.G. *et al.* (1936), *Ann. Bot.* **50**, 23−58.

[83] Mees, G.C. and Weatherley, P.E. (1957), *Proc. Roy. Soc. B.* **147**, 381–391.

[84] Milburn, J.A. and McLaughlin, M.E. (1974), *New Phytol.* **73**, 861–871.

[85] Mullins, M.G. (1970), *Ann. Bot.* **34**, 897–909.

[86] Münch, E. (1930), *Die Stoffbewegungen in der Pflanze,* Jena: Fischer.

[87] Nelson, G.D. and Gorham, P.R. (1957), *Can. J. Bot.* **35**, 339–347.

[88] Nelson, C.D. *et al.* (1961), *Plant Physiol.* **36**, 581–588.

[89] Nissen, P. (1974), *Ann. Rev. Plant Physiol.* **25**, 53–79.

[90] Parker, B.C. (1965), *J. Physiol.* **1**, 41–46.

[91] Parthasarathy, M.V. and Mühlethaler, K. (1969), *Cytobiologie,* **1**, 7–36.

[92] Pate, J.S. *et al.* (1974), *Planta,* **120**, 229–243.

[93] Peel, A.J. and Weatherley, P.E. (1959), *Nature,* **184**, 1955–1956.

[94] Penny, M.G. and Bowling, D.J.F. (1974), *Planta,* **119**, 17–25.

[95] Philips, D.A. and Cleland, C.F. (1972), *Planta,* **102**, 173–178.

[96] Pitman, M.G. (1972), *Aust. J. Biol. Sci.* **25**, 243–257.

[97] Pitman, M.G. (1975), in [11] p. 267.

[98] Postlethwait, S.N. and Rogers, B. (1958), *Am. J. Botany,* **45**, 753–757.

[99] Preston, R.D. (1958) in Everett, D.H. and Stone, F.S. (eds) *The Structure and Properties of Porous Materials,* p. 366 Butterworths, London.

[100] Qureshi, F.A. and Spanner, D.C. (1973), *J. Exptl. Botany,* **24**, 751–762.

[101] Raschke, K. and Fellows, M.P. (1971), *Planta,* **101**, 296–316.

[102] Raschke, K. and Humble, G.D. (1973), *Planta,* **115**, 47–57.

[103] Raven, J.A. (1967), *J. Gen. Physiol.* **50**, 1607–25.

[104] Raven, J.A. (1975), in [11] p. 125.

[105] Robards, A.W. and Robb, M.E. (1972), *Science,* **178**, 980–982.

[106] Salter, R.L. *et al.* (1970), *Am. J. Botany,* **57**, 916–926.

[107] Schumacher, W. (1930), *Jahrb. Wiss. Bot.* **73**, 770–823.

[108] Shone, M.G.T. *et al.* (1969), *Planta,* **86**, 301–314.

[109] Siddiqui, A.W. and Spanner, D.C. (1970), *Planta,* **91**, 181–189.

[110] Slatyer, R.O. (1967), *Plant-Water Relationships,* Academic Press, New York.

[111] Smith, F.A. (1970), *New Phytol,* **69**, 903–917.

[112] Smith, R.C. and Epstein, E. (1964), *Plant Physiol,* **39**, 338–341.

[113] Spanner, D.C. (1958), *J. Exptl. Botany,* **9**, 332–342.

[114] Spanner, D.C. and Jones, R.L. (1970), *Planta,* **92**, 64–72.

[115] Tamulevich, S.R. and Evert, R.F. (1966), *Planta,* **69**, 319–337.

[116] Thaine, R. (1962), *J. Exptl. Botany,* **13**, 152–160.

[117] Thellier, M. (1973), in [4] p. 47.

[118] Thomas, D.A. (1975), in [11] p. 377.

[119] Thompson, R.G. and Thompson, A.D. (1973), *Can J. Botany,* **51**, 933–936.

[120] Trip, P. and Gorham, P.R. (1968), *Plant Physiol.* **43**, 877–882.

[121] Tukey, H.B. Jr. (1970), *Ann. Rev. Plant. Physiol.* **21**, 305–324.

[122] Tyree, M.T. (1970), *J. Theor. Biol.* **26**, 181–214.

[123] Tyree, M.T. and Fensom, D.S. (1970), *J. Exptl. Botany,* **21**, 304–324.

[124] Van Die, J. and Tammes, P.M.L. (1975), in [142] p. 196.

[125] Vernon, L.P. and Aronoff, S. (1952), *Arch. Biochem. Biophys.* **36**, 383–398.

[126] Weatherley, P.E. and Johnson, R.P.C. (1968), *Int. Rev. Cytol.* **24**, 149–192.

[127] Weatherley, P.E. *et al.* (1959), *J. Exptl. Botany,* **10**, 1–16.

[128] Williams, S.E. and Spanswick, R.M. (1976), *J. comp. Physiol.* **108**, 211–223.

[129] Williamson, R.E. (1972), *Planta,* **106**, 149–157.

[130] Winter, K. and Lüttge, U. (1976), *Aust. J. Plant Physiol.* **3**, 653–66

[131] Wooding, F.P.B. (1971), *Phloem,* Oxford Biology Readers, Oxford University Press.

[132] Wyn Jones, R.G. (1975), in [11] p. 193.

[133] Yeo, A.R. *et al.* (1977), *J. Exptl. Botany,* **28**, 17–29.

[134] Yu, G.H. and Kramer, P.J. (1969), *Plant Physiol.* **44**, 1095–1100.

[135] Ziegler, H. (1974), *Symp. Soc. Exptl. Biol.* **28**, 43–62.

[136] Ziegler, H. (1975), in [142] p. 59.

[137] Ziegler, H. and Kluge, M. (1962), *Planta,* **58**, 144–153.

[138] Zimmermann, M.H. (1958), *Plant Physiol.* **33**, 213–217.

[139] Zimmermann, M.H. (1963), *Scient. Am.* **208**, March, 132–142.

[140] Zimmermann, M.H. (1964), *Plant Physiol.* **39**, 568–572.

[141] Zimmermann, M.H. (1969), in Wilkins, M.B. (ed) *Physiology of Plant Growth and Development,* p. 383, McGraw-Hill, London.

[142] Zimmermann, M.H. and Milburn, J.A. (eds) (1975), *Encyclopedia of Plant Physiology* N.S. Volume 1, Springer-Verlag, Berlin, Heidelberg, New York.

Index

77

79

Sodium (*continued*)
  fluxes, 10, 15−18
  levels in various materials,
    8, 19, 24, 34, 35, 37, 47
    50
  permeability coefficient, 11
  transport in symplasm, 28
Sorbitol, 47, 48
Sources of translocates, 54, 57,
    58, 60, 66, 67
Soybean, 52, 54
Specific mass transfer (SMT),
    54−56
Stachyose, 47−49, 62
Stele, 25
Stomata, 18−21
Suberin, 27
Sucrose, 47−49, 55, 61, 62, 68
Sugar cane, 50, 55, 59−61
Sulphate
  level in phloem sap, 47, 50
  level in xylem sap, 35
Sulphur, 37
Symplast, 23−31, 37, 38, 58
System I and II ion uptake, 14,
    15

Tensile strength of water, 64
Tonoplast, 14, 15, 24, 41, 42,
    45
Transcellular strands, 42, 43, 46
Transfer cells, 30−32, 36
Translocation rates, 54−56

Transpiration stream, 26, 30,
    63−66
Triticum aestivum, 55

Unloading of translocates, 58, 59
Ussing-Teorell equation, 10, 15

Verbascose, 47, 48
Vicia faba, 20
Viscosity, 54, 63, 67
Vitis vinifera, see grape
Volume flux, *see* water flux

Water flux (volume flux),
    28−30, 63
Water free space (WFS), 23
Whale, 71
White ash, 62
Willow, *see Salix*

Xanthium, 52
Xylem, fibres, 32, 33
  parenchyma, 25, 30, 32−34,
    38
  sap composition, 34, 35
  structure, 32−34
  tracheids, 32−33
  vessel elements, 32, 33

Yucca, 46, 51

Zea mays (Maize), 21, 24, 30, 34
Zinc, 51